Acclaim for **ERIK LARSON**'s

ISAAC'S STORM

"Masterful.... A thoroughly engrossing account of the catastophe."
—*The News & Observer* (Raleigh)

"A terrifying account of the storm's wrath."
—*The Seattle Times/Post-Intelligencer*

"Larson's vivid detail and storytelling ability go beyond our fascination with bizarre weather and show how natural disasters can change the course of history." —*The Hartford Courant*

"Richly imagined and prodigiously researched, [*Isaac's Storm*] pulls readers into the eye of the hurricane." —*The New York Times*

"This brilliant exploration of the hurricane's deadly force is set against the human drama of Isaac Monroe Cline.... Long after you lift your eyes from the final page, this book will bring you back to its portraits of a city under siege, and the storm's survivors and victims."
—*The Times-Picayune*

"Drawing from public records and the personal accounts of survivors, Larson tracks in vivid detail both the path of the hurricane and the trajectory of Isaac Cline's carrer." —*The Atlanta Journal-Constitution*

"Larson offers Dantesque images of trees, street lamps, houses and furnishings being turned into projectiles . . . of wind-whipped walls of water pressing the life out of entire city blocks." —*The Plain Dealer*

"In all the books about disasters, few have assembled so many nuanced details into the kind of flood that Larson releases when the storm surges." —*San Francisco Chronicle*

Also by ERIK LARSON

The Naked Consumer

Lethal Passage

ERIK LARSON

ISAAC'S STORM

Erik Larson, a contributor to *Time* magazine, is the author
of *The Naked Consumer* and *Lethal Passage*. His work has
appeared in *The Atlantic Monthly, Harper's, The New Yorker,*
and other national magazines. He lives in Seattle.

ISAAC'S STORM

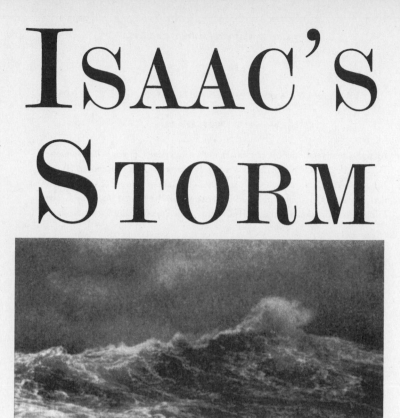

A Man, a Time, and the Deadliest Hurricane in History

ERIK LARSON

VINTAGE BOOKS

A DIVISION OF RANDOM HOUSE, INC.   NEW YORK

For Chris, Kristen, Lauren, and Erin

FIRST VINTAGE BOOKS EDITION, JULY 2000

Copyright © 1999 by Erik Larson

All rights reserved under International and Pan-American Copyright Conventions. Published in the United States by Vintage Books, a division of Random House, Inc., New York and simultaneously in Canada by Random House of Canada Limited, Toronto. Originally published in hardcover in the United States by Crown Publishers, a division of Random House, Inc., New York, in 1999.

Vintage and colophon are registered trademarks of Random House, Inc.

The map on pp. x-xi is being used by courtesy of the Rosenberg Library, Galveston, Texas.

The Library of Congress has cataloged the Crown edition as follows:
Larson, Erik.
Isaac's storm : a man, a time, and the deadliest hurricane in history / Erik Larson.
— 1st ed.
p. cm.
Based on the diaries of Isaac Monroe Cline and on contemporary accounts.
1. Galveston (Tex.)—History—20th century. 2. Hurricanes—Texas—Galveston—History—20th century. 3. Floods—Texas—Galveston—History—20th century. 4. Cline, Isaac Monroe. 5. Galveston (Tex.) Biography.
I. Cline, Isaac Monroe. II. Title.
F394.G2L37 1999
976.4'139—dc21 99-25515
CIP
ISBN 0-609-60233-0

Vintage ISBN: 0-375-70827-8

Author photograph © Roseanne Olson
Book design by Leonard Henderson

www.vintagebooks.com

Printed in the United States of America
10

Contents

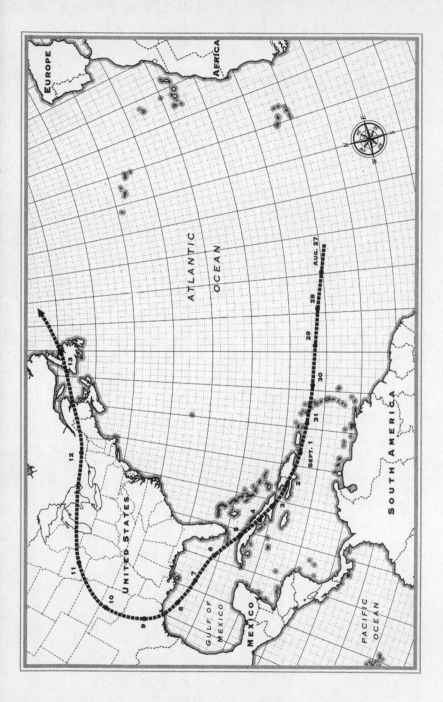

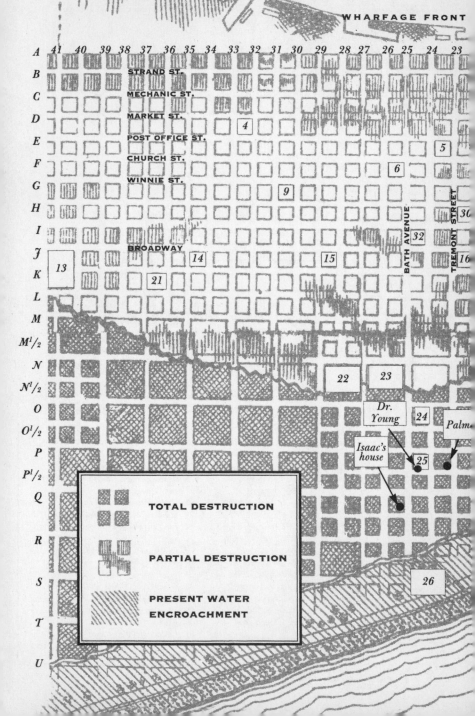

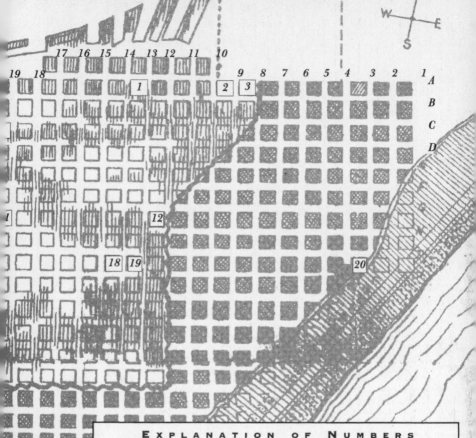

EXPLANATION OF NUMBERS

1. Elevator A, partly destroyed.
2. Medical College, slightly damaged.
3. Sealy Hospital, partly destroyed.
4. City Gas Works, almost totally ruined.
5. Tremont Hotel, badly damaged.
6. Custom House, slightly damaged.
7. Harmony Hall, badly damaged.
8. City Hall, almost totally destroyed.
9. City Water Works, significant damage.
10. Ball High School. badly damaged.
11. Court House, very little damage.
12. Rosenberg School, badly damaged.
13. Episcopal Cemetery, almost ruined.
14. St. Patrick Church, complete wreck.
15. First Baptist Church. demolished.
16. German Catholic Church, little damage.
17. City Street Railway power house, ruined.
18. Gresham's residence, no damage.

19. St. Mary's University for Boys, badly damaged. Church of the Sacred Heart, completely destroyed.
20. Lucas Terrace, a complete wreck.
21. Grace Episcopal Church, very little, if any, damage.
22. Garten Verein, small damage.
23. Ursuline Convent, badly damaged.
24. Rosenberg Women's Home, badly damaged.
25. Bath Avenue Public School, almost wrecked.
26. Beach Hotel.
27. Grand Opera House, badly damaged.
28. Artillery Hall, some damage.
29. Synagogue, little damage.
30. Baptist Church, small damage.
31. Cathedral, little damage.
32. George Sealy's residence, little damage.

GULF OF MEXICO

TELEGRAM

Washington, D.C.

Sept. 9, 1900

To: Manager, Western Union

Houston, Texas

Do you hear anything about Galveston?

Willis L. Moore,
Chief, U.S. Weather Bureau

THE BEACH

September 8, 1900

THROUGHOUT THE NIGHT of Friday, September 7, 1900, Isaac Monroe Cline found himself waking to a persistent sense of something gone wrong. It was the kind of feeling parents often experienced and one that no doubt had come to him when each of his three daughters was a baby. Each would cry, of course, and often for astounding lengths of time, tearing a seam not just through the Cline house but also, in that day of open windows and unlocked doors, through the dew-sequined peace of his entire neighborhood. On some nights, however, the children cried only long enough to wake him, and he would lie there heart-struck, wondering what had brought him back to the world at such an unaccustomed hour. Tonight that feeling returned.

Most other nights, Isaac slept soundly. He was a creature of the last turning of the centuries when sleep seemed to come more easily. Things were clear to him. He was loyal, a believer in dignity, honor, and effort. He taught Sunday school. He paid cash, a fact noted in a directory published by the Giles Mercantile Agency and meant to be held in strictest confidence. The small red book fit into a vest pocket and listed nearly all Galveston's established citizens—its police officers, bankers, waiters, clerics, tobacconists, undertakers, tycoons, and shipping agents—and rated them for credit-worthiness, basing this appraisal on secret reports filed anonymously by friends and enemies. An asterisk beside a name meant trouble, "Inquire at Office," and marred the fiscal reputations of such people as Joe Amando, tamale vendor; Noah Allen, attorney; Ida Cherry, widow; and August Rollfing, housepainter. Isaac Cline got the highest rating, a "B," for "Pays Well, Worthy of Credit." In November of 1893, two years after Isaac arrived in Galveston to open the Texas Section of the new U.S. Weather Bureau, a government inspector wrote: "I suppose there is not a man in the Service on Station Duty who does more real work than he. . . . He takes a remarkable degree of interest in his work, and has a great pride in making his station one of the best and most important in the country, as it is now."

Upon first meeting Isaac, men found him to be modest and self-effacing, but those who came to know him well saw a hardness and confidence that verged on conceit. A New Orleans photographer captured this aspect in a photograph that is so good, with so much attention to the geometries of composition and light, it could be a portrait in oil. The background is black; Isaac's suit is black. His shirt is the color of bleached bone. He has a mustache and goatee and wears a straw hat, not the rigid cake-plate variety, but one with a sweeping scimitar brim that imparts to him the look of a French painter or riverboat gambler. A darkness suffuses the photograph. The brim shadows the top of his face. His eyes gleam from the darkness. Most striking is the careful positioning of his hands. His right rests in his lap, gripping what could be a pair of gloves. His left is positioned in midair so that the diamond on his pinkie sparks with the intensity of a star.

There is a secret embedded in this photograph. For now, however, suffice it to say the portrait suggests vanity, that Isaac was aware of himself and how he moved through the day, and saw himself as something bigger than a mere recorder of rainfall and temperature. He was a scientist, not some farmer who gauged the weather by aches in a rheumatoid knee. Isaac personally had encountered and explained some of the strangest atmospheric phenomena a weatherman could ever hope to experience, but also had read the works of the most celebrated meteorologists and physical geographers of the nineteenth century, men like Henry Piddington, Matthew Fontaine Maury, William Redfield, and James Espy, and he had followed their celebrated hunt for the Law of Storms. He believed deeply that he understood it all.

He lived in a big time, astride the changing centuries. The frontier was still a living, vivid thing, with Buffalo Bill Cody touring his Wild West Show to sellout crowds around the globe, Bat Masterson a sportswriter in New Jersey, and Frank James opening the family ranch for tours at fifty cents a head. But a new America was emerging, one with big and global aspirations. Teddy Roosevelt, flanked by his Rough Riders, campaigned for the vice presidency. U.S. warships steamed to quell the Boxers. There was fabulous talk of a great American-built canal that would link the Atlantic to the Pacific, a task at which Vicomte de Lesseps and the French had so catastrophically failed. The nation in 1900 was swollen with pride and technological confidence. It was a time, wrote Sen. Chauncey Depew, one of the most prominent politicians of the age, when the average American felt "four-hundred-percent bigger" than the year before.

There was talk even of controlling the weather—of subduing hail with cannon blasts and igniting forest fires to bring rain.

In this new age, nature itself seemed no great obstacle.

ISAAC'S WIFE, CORA, lay beside him. She was pregnant with their fourth child and the pregnancy had entered a difficult stretch, but now she slept peacefully, her abdomen a pale island against the darkness.

The heat no doubt contributed to Isaac's sleeplessness. It had been a problem all week, in fact all summer, especially for Cora, whose pregnancy had transformed her body into a furnace. Temperatures in Galveston had risen steadily since Tuesday. The heat broke 90 degrees on Thursday, and hit 90 again on Friday. Moisture from weeks of heavy rain concentrated in the air until the humidity was unbearable. Earlier that week Isaac had read in the Galveston *News* how a heat wave in Chicago had killed at least three people. Even the northernmost latitudes were experiencing unusual levels of warmth. For the first time in recorded history, the Bering Glacier in what eventually would become Alaska had begun to shrink, sprouting rivers, calving icebergs, and ultimately shedding six hundred feet of its depth. A correspondent for *The Western World* magazine wrote, "The summer of 1900 will be long remembered as one of the most remarkable for sustained high temperature that has been experienced for almost a generation."

The prolonged heat had warmed the waters of the Gulf to the temperature of a bath, a not-unhappy condition for the thousands of new immigrants just arrived from Europe at the Port of Galveston, known to many as the Western Ellis Island. Some camped now on the beach near the Army's new gun emplacements, steeling themselves for the long journey north to open land and the riches promised them by railroads intent on populating America's vast undeveloped prairie. In a pamphlet called *Home Seekers,* the Atchison, Topeka and Santa Fe described the lush land of the Texas coast as "waiting to be tickled into a laughing harvest." The railroad come-ons painted Texas as a paradise of benign weather, when in fact hurricanes scoured its coast, plumes of hot wind baked apples in its trees, and "blue northers" could drop the temperature fifty degrees in a matter of minutes. To Isaac, such quirks of weather were a fascination, and not just because he happened to be the chief weatherman in Texas. He was also a physician. He no longer saw patients, but had become a pioneer in medical climatology, the study of how weather affects people, and in this carried forth a tradition laid down by Hippocrates, who believed climate determined the character of men and nations.

Hippocrates advised any physician arriving in an unfamiliar town to first "examine its position with respect to the winds."

. . .

As FRIDAY NIGHT ebbed into Saturday, the air at last cooled. The sudden change in temperature would come as a delightful surprise to others in Galveston, but to Isaac it was one more flicker of trouble.

He let his mind wander through the house. He heard no sound from the children's bedrooms. His eldest daughter, Allie May, was now twelve; his middle daughter, Rosemary, was eleven. His youngest, Esther Bellew, was six, but he still called her his baby. He heard nothing also from his brother, Joseph, who lived in the house. Eight years earlier, Joseph had come to work for Isaac as an assistant observer. The two men were still close, but soon any tie between them would be severed for all time and each would pass the remainder of his life as if the other never existed. Joseph was twenty-nine. Isaac was thirty-eight.

Isaac's house stood at 2511 Avenue Q, just three blocks north of the Gulf. It was four years old and replaced a previous house that had burned in a fire in November 1896. Isaac had ordered this house built atop a forest of stilts with the explicit goal of making it impervious to the worst storms the Gulf could deliver. It had two stories, with porches or "galleries" off each floor in the front and rear, and a small building in the backyard that served as a stable. The house was ideally situated. On Sundays Isaac and his family would join the torrent of other families walking down 25th Street toward the big Victorian bathhouses built over the Gulf. Sometimes they walked to Murdoch's; other days they chose the Pagoda Company Bath House, with its two large octagonal pavilions and sloping pagoda roofs. The Clines reached it by walking the length of a 250-foot boardwalk that began at the foot of 24th Street, rose 16 vertical feet above the beach, and ran another 110 feet out over the waves, as if its builders believed they had conquered the sea for once and for all. An electric wire ran to a pole far out in the surf, where it powered a lamp suspended over the water. At night bathers gathered like insects.

Isaac heard the usual sounds that sleeping houses make, even houses as strong as his. He heard the creaking and sighing of beams, posts, and joists as the relatively new lumber of his home absorbed the moisture of the night and released the last heat of day. He heard the susurrus of curtains luffed by the breeze. There would have been mice, too, and mosquitoes. If people sought to protect themselves at all, they propped tents of fine, gauzelike netting over their beds. No one had window screens.

As Isaac listened, background noises came forward. One noise in particular. It was more than noise, really. If Isaac lay very still, he could feel the shock waves climb the stilts of his house, the same way he felt the vibration of the pipe organ Cora played at church each Sunday. To children in houses all along the beach, particularly the ninety-three children in the big, sad St. Mary's Orphanage two miles west at the very edge of the sea, the sound was a delight. They heard it and felt it and dreamed it. To some, each shock wave was the concussion of British artillery in the Boer War or a ghost gun from the dead *Maine,* or perhaps the thud of an approaching giant. A welcome giant. The shuddering ground promised a delightful departure from the steamy sameness of Galveston's summers, and it came with exquisite timing: Saturday. Only hours ahead lay Saturday night, the most delicious night of all.

But the sound frightened Isaac. The thudding, he knew, was caused by great deep-ocean swells falling upon the beach. Most days the Gulf was as placid as a big lake, with surf that did not crash but rather wore itself away on the sand. The first swells had arrived Friday. Now the booming was louder and heavier, each concussion more profound.

ISAAC WOKE AGAIN at 4:00 A.M., but this time the cause was obvious. His brother stood outside the bedroom door tapping gently and calling his name.

Joseph too had been unable to sleep. Not a terribly creative man, he described this feeling as a sense of "impending disaster." He had stayed up until midnight recording weather observations from a bank of instruments mounted on the roof of the Levy Building, a four-story brick building in the heart of Galveston's commercial district. The barometers had captured only a slight decrease in pressure. The anemometer, which caught the wind in cups mounted at opposite ends of crossed metal bars, recorded wind speeds of eleven to nineteen miles an hour. It was capable of measuring velocities as high as one hundred miles an hour, but conditions had never come close to testing this capacity, nor did any rational soul believe they ever would. Throughout Friday afternoon and evening, a peculiar oppressiveness had settled over the city. Temperatures remained high well into the night.

None of these observations was enough by itself to raise concern. For days, however, Isaac had been receiving cables from the Weather Bureau's Central Office in Washington describing a storm apparently of tropical origin that had drenched Cuba. Although Isaac did not know it, there was confusion about the storm's true course, debate as to its character. The bureau's men in Cuba said the storm was nothing to worry about; Cuba's own weather observers, who had pioneered hurricane detection, disagreed. Conflict between both groups had grown increasingly intense, an effect of the unending campaign of Willis Moore, chief of the U.S. Weather Bureau, to exert ever more centralized control over forecasting and the issuance of storm warnings. The bureau had long banned the use of the word *tornado* because it induced panic, and panic brought criticism, something the bureau could ill afford. Earlier that week, Moore had sent Galveston a telegram asserting yet again that only headquarters could issue storm warnings.

At 11:30 A.M. on Friday, Moore had sent another telegram, this one notifying Isaac and other observers of a tropical storm centered in the Gulf of Mexico south of Louisiana, "moving slowly northwest." The telegram predicted "high northerly winds tonight and Saturday with probably heavy rain."

Again, nothing especially worrisome. Tropical storms came ashore every summer. They brought wind and rain, even some flooding. Damage was rare. No one got hurt. But in one respect the telegram did surprise Isaac. Until now, Moore's cables had expressed absolute confidence the storm was moving north toward the Atlantic coast.

Isaac got out of bed, careful not to wake Cora. Joseph's intrusion annoyed him. There was tension between the brothers. Nothing open— at least not yet. Just a persistent low-grade rivalry.

He and Joseph descended to the kitchen, careful to avoid waking the children, and there by sheer force of habit Isaac put on a pot of coffee. They talked about the weather. A familiar dynamic emerged. Joseph, as the younger brother and junior employee eager to prove himself, made the case too strongly that something peculiar was happening and that Washington *must* be informed. Isaac, ever confident, told Joseph to get some sleep, that he would take over and assess the situation and if necessary telegraph his findings to headquarters.

Isaac dressed. He stepped out onto the first-floor porch. With most of the block that faced him across Avenue Q still undeveloped, he had an unobstructed view of the sky and the cityscape to the north. He saw lime-washed bungalows and elaborate three-story homes with gables, bays, and cupolas, and just beyond these the big Rosenberg Women's Home and the Bath Avenue Public School. At the corner, to his right and across the street, stood the three-story home of the Neville family, windows open, dew and drizzle darkening its intricate slate roof. Ever since the great fire of 1885, Galveston had required that roofs be shingled with slate instead of wood as a safety precaution, but in just a few hours the shingles from the Neville house, Isaac's house, and thousands of others throughout Galveston would begin whirling through the air with an effect that evoked for many older residents the gore-filled afternoons they spent at Chancellorsville and Antietam.

Isaac harnessed his horse to a small two-wheeled sulky that he used mostly when hunting and with a gentle click of the reins set out for the beach three blocks south.

IT WAS A gorgeous morning, the breeze soft and suffused with mist, jasmine, and oleander. Stratus and cumulus clouds filled most of the sky, some bellying almost to the sea, but Isaac also saw patches of dawn blue rimmed with cloudsmoke. To his left, behind the clouds, the sun had begun to rise and at odd moments it turned the clouds orange-gray, like fire behind smoke. Seagulls hung in threes at fixed points in the sky where they rode head-on into the unaccustomed north wind, wing tips flinching for purchase. The wheels of Isaac's sulky broadcast a reassuring crunch as they moved over the pavement of crushed oyster shells.

By now the most industrious children were rising to do their chores and get them out of the way so they could go to the beach as early as possible. Everyone reveled in the refreshing coolness. Rabbi Henry Cohen was awake and preparing for Saturday's services. Dr. Samuel O. Young, an amateur meteorologist and secretary of the Galveston Cotton Exchange, was having breakfast and planning his own early-morning trip to the beach. At 18th Street and Avenue O½, in a small two-story rental house, Louisa Rollfing made breakfast for her husband, August, who was due downtown that morning to continue the painting of a commercial building. Louisa looked out the window and as always felt just a hint of disappointment, or maybe sorrow, for although she liked Galveston, she still was not used to the landscape. To her, palms and live oak did not qualify as trees. She missed the great green-black forests of her childhood home in Germany with trees "so old and large, that in some places it is almost dark in daytime."

Visitors approaching Galveston from the sea saw it as a brilliant swath of light between sea and sky, like mercury floating on a deep blue plain. In the summer of 1900, a boy named John W. Thomason Jr.—later to become a well-known writer of military history—arrived to spend his vacation with his grandfather in a cottage off Broadway, half a dozen blocks from Isaac Cline's office. "The Gulf breeze cooled the city at nightfall; one of the most beautiful beaches in the world offered delightful surf-bathing; and you saw everybody there in the afternoons, bathing, promenading or driving in carriages on the smooth, crisp sands." He left town on Saturday, September 1, exactly a week before Isaac's trip to the beach, very sad to leave. He looked back with longing as his train clicked over the long wooden trestle to the mainland and his newfound friends receded into the steam rising from Galveston Bay. "That city as it was," he wrote, "I never saw again, nor some of the boys and girls I knew there."

Where critics most faulted Galveston was for its lack of geophysical presence. The city occupied a long, narrow island that also formed the southern boundary of Galveston Bay, spanned by three railroad trestles and a wagon bridge. Its highest point, on Broadway, was 8.7 feet above sea level; its average altitude was half that, so low that with each one-foot increase in tide, the city lost a thousand feet of beach. Josiah Gregg, one of America's most celebrated traveler-raconteurs, wrote in his diary in November 1841 of hearing about a past flood in which "this island was so completely overflowed that a small vessel actually sailed out over the middle of it." He did not believe the story. He could see, however, that someday flooding might "even endanger lives."

Regardless of one's view, the fact was that Galveston in 1900 stood on the verge of greatness. If things continued as they were, Galveston soon would achieve the stature of New Orleans, Baltimore, or San Francisco. The New York *Herald* had already dubbed the city the New York of the Gulf. But city leaders also knew there was only room on the Texas coast for one great city, and that they were in a winner-take-all race against Houston, just fifty miles to the north. As of 1900, Galveston had the lead. The year before, it had become the biggest cotton port in the country and the third-busiest port overall. Forty-five steamship lines served the city, among them the White Star Line, which provided service between Galveston and Europe and in just over a decade would lose a great ship to hubris and ice. Consulates in the city represented sixteen countries, including Russia and Japan. And Galveston's population was growing fast. On Friday, September 7, Isaac had read in the *News* the first brief report on the Galveston count of the 1900 census, which found that the city had grown 30 percent in only ten years.

Galveston now had electric streetcars, electric lights, local and long-distance telephone service, two domestic telegraph companies, three big concert halls, and twenty hotels, the most elegant being the Tremont, south of Isaac's office, with two hundred ocean-facing rooms, fifty "elegant" rooms with private baths, and its own electric-power plant.

What most marked the city was money. As early as 1857 Galveston had achieved a reputation as a cosmopolitan town with a passion for fine things. One of its French chefs distinguished himself with a fusion of frontier and Continental cuisine that featured "beefsteak goddam a la mode." By 1900, the city was reputed to have more millionaires per square mile than Newport, Rhode Island. Much of their money was vividly on display in the ornate mansions and lush gardens of Broadway, the city's premier street.

The city offered everything from sex to sacks of Tidal Wave Flour. For the grieving rich, the giant livery and funeral works of J. Levy and Brothers offered a very special option: "A child's white hearse and harness, with white horses."

WINDOWS WERE OPEN in all the houses Isaac passed, and this imparted to the city an aura of vulnerability. Suddenly the noise of the sulky's wheels seemed more jarring than reassuring. Ordinarily the great bathhouses at the end of the street would have brightened Isaac's mood, but today they looked swollen and worn; they floated on cushions of greenish mist like castles from the mind of Poe.

Isaac drove until he had a clear view of the Gulf, then stopped the sulky. He stood, pulled out his watch, and began timing the long hills of water that rolled toward the beach. The crests of the waves were brown with sand, but on the surface between crests the spindrift laid intricate patterns of shocking-white lace.

Isaac knew the low-pressure center of the storm had to be somewhere off to his left, out in the Gulf. It was a fundamental tenet of marine navigation, one he explained during a lecture at the Galveston YMCA on a Saturday evening in 1891. Large crowds gathered for such talks. They consumed the spoken word the way later men would consume television. In the northern hemisphere, Isaac told his audience, the winds of tropical cyclones always move counterclockwise around a central area of low pressure. "Stand with your back to the wind," he said, "and the barometer will be lower on your left than on your right."

The swells came very slowly, at intervals of one to five minutes. To lay observers, this slow pace might have seemed reassuring. In fact, the slowness made the swells far more ominous, a principle Isaac only vaguely understood. Many years later he would write, "If we had known then what we know now of these swells, and the tides they create, we would have known earlier the terrors of the storm which these swells . . . told us in unerring language was coming."

ISAAC TURNED HIS sulky around and headed back toward his office. The breeze was now head-on and ruffled the mane of his horse. The oyster-shell paving gave way to heavy wooden blocks and these imparted to the sulky a beat like that of a swiftly moving train. The north wind brought Isaac the perfume of a waking city: the clean, almost minty, smell of freshly cut lumber from the Hildenbrand planing mill; coffee from the bulk roasters in the alley between Mechanic and Market; and always, everywhere, the scent of horses.

At the Levy Building, Isaac walked the three flights to the bureau, stopped inside for a moment, then continued up to the roof. To the east and south he saw the sea; to the west, the spires of St. Patrick's Church, still under construction. The bureau's storm flag, a single crimson square with a smaller black square at its center, rippled from a tower.

The barometer showed that atmospheric pressure had fallen only slightly from the night before. "Only one-tenth of an inch lower," Isaac said.

Nothing in the sky, the instruments, or the cables from Washington indicated a storm of much intensity. "The usual signs which herald the approach of hurricanes were not present in this case," he said. "The brick-dust sky was not in evidence in the smallest degree."

Even so, the day felt wrong. Ordinarily, offshore winds kept the surf and tides down, but now, despite the brisk north wind, both the surf and tide were rising. It was a pattern new to Isaac.

He drove his sulky back to the beach. He again timed the swells. He noted their shape, their color, the arc they produced as they mounted the sand. They were heavier now and pushed seawater onto the streets closest to the beach.

Isaac returned to his office and composed a telegram to the Central Office in Washington. He ended the telegram: "Such high water with opposing winds never observed previously."

Isaac's concern was tempered by his belief that no storm could do serious damage to Galveston. He had concluded this on the basis of his own analysis of the unique geography of the Gulf and how it shaped the region's weather. In 1891, in the wake of a tropical storm that Galveston weathered handily, the editors of the Galveston *News* invited Isaac to appraise the city's vulnerability to extreme weather. Isaac, father of three, husband, lover, scientist, and creature of the new heroic American age, wrote: "The opinion held by some who are unacquainted with the actual conditions of things, that Galveston will at some time be seriously damaged by some such disturbance, is simply an absurd delusion."

At the top of the Levy Building the anemometer spun. The wind vane shifted ever so slightly. The self-recording barometer etched another tiny decline.

FAR OUT TO sea, one hundred miles from where Isaac stood, Capt. J. W. Simmons, master of the steamship *Pensacola,* prayed softly to himself as horizontal spheres of rain exploded against the bridge with such force they luminesced in a billion pinpoints of light, like fireworks in a green-black sky.

He had stumbled into the deadliest storm ever to target America. Within the next twenty-four hours, eight thousand men, women, and children in the city of Galveston would lose their lives. The city itself would lose its future. Isaac would suffer an unbearable loss. And he would wonder always if some of the blame did not belong to him.

This is the story of Isaac and his time in America, the last turning of the centuries, when the hubris of men led them to believe they could disregard even nature itself.

PART I

The Law of Storms

_____

THE STORM

Somewhere, a Butterfly

IT BEGAN, AS all things must, with an awakening of molecules. The sun rose over the African highlands east of Cameroon and warmed grasslands, forests, lakes, and rivers, and the men and creatures that moved and breathed among them; it warmed their exhalations and caused these to rise upward as a great plume of carbon, oxygen, nitrogen, and hydrogen, the earth's soul. The air contained water: haze, steam, vapor; the stench of day-old kill and the greetings of men glad to awaken from the cool mystery of night. There was cordite, ether, urine, dung. Coffee. Bacon. Sweat. An invisible paisley of plumes and counterplumes formed above the earth, the pattern as ephemeral as the copper and bronze veils that appear when water enters whiskey.

Winds converged. A big, hot easterly raced around a heat-induced low in the Sahara, where temperatures averaged 113 degrees Fahrenheit, heat scalded the air, and winds filled the sky with dust. This easterly blew toward the moist and far cooler bulge of West Africa. High over the lush lands north of the Gulf of Guinea, over Ouagadougou, Zungeru, and Yamoussoukro, this thermal stream encountered moist monsoon air blowing in from the sea from the southwest. The monsoon crossed the point where zero latitude and zero longitude meet, and entered the continent over Nigeria.

Where these winds collided, they produced a zone of instability. The air began to undulate.

THE SEAS WERE hot. The land was hot. Throughout much of the United States temperatures rose into the nineties and often broke 100. Heat suffused the Rockies, Nebraska, Kansas, Missouri, Oklahoma, and a vast swath of country from the Gulf all the way to Pennsylvania. At 3:00 P.M. on Saturday, August 11, the temperature in Philadelphia hit 100.6 degrees. There was no air-conditioning. Trains were hot. Suits were black wool. Dresses were taffeta, mohair, gabardine. Carriages had black canvas tops, black-enameled bodies. Passengers roasted. Horses glistened. That same Saturday, thirty people in New York City died of heat prostration. Three children died when they fell from fire escapes where they had hoped to find a breeze. A high-pressure zone stretched from the Midwest far into the Atlantic and halted the flow of air over much of the nation. There was no breeze to find. "The air near the surface of the earth became superheated," wrote Prof. E. B. Garriott, the Weather Bureau's chief forecaster at the time. "Considered as a whole, the month of August, 1900, was the warmest August on record generally from the upper Mississippi Valley over the Lake region, Ohio Valley, and Middle Atlantic States."

Which meant the heat embraced most of the nation's population. Everyone shared in the suffering. What made the heat wave exceptional was not the maximum temperature recorded from city to city, but the sheer persistence of the heat. Springfield, Illinois, reported the longest hot spell in twenty years: twelve consecutive days with temperatures of 90 or higher. The men at Weather Bureau headquarters suffered deeply as the mercury hit or surpassed 96 degrees seven days in a row. In August, mean temperatures in Albany, Atlantic City, Baltimore, Chicago, Cincinnati, Erie, New York, and Philadelphia were the highest they had been since the bureau began keeping formal records in 1873.

In Galveston there was heat *and* rain. From mid-July to mid-August, a succession of tropical squalls swept from the Gulf and deluged Galveston. In one twenty-four-hour period, the city got fourteen inches of rain. Some streets flooded. Little boys converted tubs to boats and sailed downtown. A horse drowned. Total rainfall for that storm alone was sixteen inches in forty-eight hours, five inches greater than Galveston's previous record set in September 1875 when a hurricane struck Indianola on Matagorda Bay, 150 miles down the Texas coast. In Paris, Texas, lightning demolished a tree. Ten billion joules of energy leaped to a porch ten feet away and knocked five children unconscious. Crickets swarmed Waco. The streets crunched. Bugs heaping under arc lights halted trolleys. Squads of citizens used unslacked lime and coal oil to drive the bugs away. The fire department deployed hoses.

The waters of the Gulf got hot.

OVER THE NIGER, the colliding winds veered and arced. Thunderstorms of great violence purpled the sky. A huge parcel of air began circling slowly, far too high for anyone on the ground to notice. The powerful Saharan wind swept it west toward the Atlantic as a wave of turbulence, thunderstorms, and driving rain.

Within this "easterly wave," moisture-freighted air rose high into the troposphere, the first layer of sky and the realm where all weather occurs. The air cooled rapidly as it pierced colder and colder layers of atmosphere and encountered lower and lower pressure. The lower the pressure, the more the air expanded. As it expanded it cooled. It continued to rise but less than a mile above the earth crossed a threshold, and a phase change occurred. The air got so cold, it could no longer retain the water it carried. The vapor condensed en masse, as if at the tap of a conductor's baton. The resulting droplets were so tiny they remained suspended in the rising air.

The updrafts pushed the droplets higher and higher at up to one hundred miles an hour. At four miles above the ground the droplets froze, and the rising air became filled with snowflakes and shards of ice. Men on the ground saw blossoms of cotton with flat gray bottoms that marked the altitude where condensation had begun. Children saw camels, rabbits, and cannon fire. The clouds bloomed before their eyes. Cells within grew and quickly expired. Some cells smoked into the sky like Christmas rockets. Others became massive Gibraltars of condensed water, *Cumulus congestus;* some rose higher, *Cumulonimbus calvus.* In the pillars that reached the top of the troposphere, temperatures fell to 100 degrees below zero. Tiny hexagonal mirrors of ice drifted from the peaks in lovely translucent veils, or "virga."

Something powerful and ultimately deadly occurred within these clouds. As the water rose and cooled and condensed, it also released heat. In the sky over Africa in August 1900, trillions upon trillions of water molecules began breathing tiny fires. This heat propelled the air even higher into the atmosphere until the cloudtops flattened to form *Cumulonimbus capillatus incus. Incus* meaning "anvil," the name too of an

anvil-shaped bone in the human ear. These were thunderheads. "Convection." Higher up, the strongest clouds penetrated the stratosphere. Soon an army of great thunderheads
was marching west along the horizon, watched closely by the
captains of British ships sailing down the African coast with
fresh troops for the Boer War. Seventy to eighty such waves
drifted from West Africa into the Atlantic every summer, some
dangerous, most not. The captains knew them less as weather,
more as geography—something to watch to fill the long hours
at sea. At dawn and dusk, the distant clouds warmed the sky
with color. Rain smudged from their bottoms in fallstreaks.
Frozen virga drifted from their glaciated tops. When the light
was just right or a squall was near, the clouds formed an
escarpment of black. Frigate birds sidelit by the sun drifted in
the foreground and flecked the sky with diamond.

Ships directly in the path of the August wave got a different
view. Each wave had a "period" of four days, meaning a ship
in a fixed location would experience a cycle of weather that
repeated itself every four days. On the first day the air was hot
and dry, like a desert at sea. No clouds, but also very little blue
sky. The only blue was directly overhead. Everywhere else the
sky was white, the horizon like milk—all of this caused by dust
carried from the deserts of Africa.

Soon, however, the sky filled with puffy clouds, *Cumulus
humilus*, the pretty fair-weather cumulus of the finest summer days. As the wave advanced, these grew fatter and taller.
High clouds arrived next, first icy cirrus, then a gray ceiling of
cirrostratus. The skies got darker, the cloud ceiling lower. A
fine drizzle began to fall. A squall line of thunderstorms followed, cousins of the great storms that just a few days earlier

had driven the shopkeepers of Dakar to seek shelter. The storms brought thunder and lightning, but were nowhere near as intense as they had been over the West African bulge. They dropped the temperature at sea level to below 70 degrees. For anyone acclimated to the humid warmth of the tropics, suddenly the air was downright cold. It was jacket weather on Cape Verde.

The squalls passed. The sky cleared. The cycle began again.

WHEREVER THE AUGUST wave traveled, it dropped the pressure exerted by the atmosphere. At first the decline was slight, but soon warm air flowed upward through the thunderheads heating the air and reducing its weight, thereby reducing the pressure it exerted on the ocean surface. The heating produced a basin of low pressure that drew air, as wind, from surrounding regions of higher pressure. Meanwhile, ambient upper-level winds whisked away the air exiting from the top of the storm. The faster the upper air departed, the faster the lower air arrived. A few clouds became so immense they began to shape the behavior of the entire mass.

The storm could have continued growing, but conditions were not quite right. The air moving from its top had begun to descend, but in a form very different from when it first entered the storm. Stripped of its moisture, this descending air was cool and dry. Cataracts of spent air fell toward the sea beyond the boundaries of the storm, but the storm's appetite had grown so large it now summoned this air as well. The cool air became caught—"entrained"—in the moist sea-level winds rushing toward the storm. As this dry air mixed with the moist, it banked the fires rising through the clouds above.

For the moment, the system stabilized.

IN GALVESTON, THE humidity was nearly one 100 percent. To move was to drip. It was too hot to put on a bathing suit. "Brown is the new color for bathing suits," the Galveston *News* reported in the caption of a photograph showing the latest in coastal chic. "This one of a rich leaf brown mohair has yoke, collar and bands of white mohair striped with black braid."

Mohair.

Every day an ad in the Galveston *News* for Dr. McLaughlin's Electric Belt asked: "Weak Men—Are You Sick?"

MOST TROPICAL DISTURBANCES dissipated over the open sea. They collided with powerful winds from the west that dipped from the middle latitudes and blew the tops off their thunderheads. They encountered pools of cold water. They entrained so much dry air they lost their passion. Their pillars of smoke and light became mist. Most of the time.

Occasionally they became killers, although exactly why remained a mystery even at the end of the twentieth century. Satellites sharpened the ability of forecasters to monitor hurricane motion but could not capture the instant of transfiguration. No matter how closely meteorologists analyzed satellite biographies of hurricanes, they still could not isolate the exact coding that destined a particular easterly wave to a future of murder and mayhem. Satellites could document changes in temperature of a few thousandths of a degree and capture features as small as a foot wide or a few centimeters tall. "But suppose," wrote Ernest Zebrowski Jr., in *Perils of a Restless Planet*, "that a tropical storm develops, and that we play back the data record of the previous few days. What do we find as we go back in time? A smaller storm, and yet a smaller disturbance, then a warm moist windy spot, then a set of atmospheric conditions that looks no diffferent from that at many other locations in the tropics."

Zebrowski proposed that the answer might lie in the science of "nonlinear dynamics": chaos theory and the famous butterfly effect. He framed the question this way: "Could a butterfly in a West African rain forest, by flitting to the left of a tree rather than to the right, possibly set into motion a chain of events that escalates into a hurricane striking coastal South Carolina a few weeks later?"

To Zebrowski, the fact that the most detailed satellite analysis could not detect a trigger suggested that tropical storms might be influenced by forces too subtle to measure. He noted that a tiny change in the variables entered into computer models of hurricane development could yield dramatic variation later on. "One simulated storm may veer northward while another continues westward, one may intensify while another is dying, or one may stand stationary while another gallops toward a shoreline."

Every hurricane, however, had characteristics similar to those of every other hurricane. Each, for example, developed thunderstorms and began to rotate. In chaos theory, these points of broadly similar behavior were "strange attractors." Subtle forces could launch a system from one attractor to another—a chance gust of wind, a plume of hot sea, maybe even the sudden burst of heat from a British frigate during a gunnery drill off Dakar.

"Add a little glitch, a metaphorical butterfly, to a complex process," Zebrowski wrote, "and sometimes you get an outcome no rational person would ever have expected."

AS GALVESTON STEAMED, the world seethed. The Boxer Rebellion intensified. The British public grew weary of the Boer War. When Boer snipers fired on a British troop train, a British general ordered every house within ten miles burned to the ground. The order shocked London. A madman assassinated Italy's King Humbert. In Paris, another assassin tried to kill the shah of Persia. Bubonic plague turned up in London and Glasgow. William Jennings Bryan stumped for the presidency and railed at America's new imperialist bent, in particular the widely held belief that expansion overseas was America's destiny. "Destiny," he thundered, "is the subterfuge of the invertebrate. . . ."

The speech ran on for eight thousand words. Despite the heat, the house was packed.

THE SEAS WERE busy. A few ships must have encountered the thunder and rain but apparently their crews did not see it as anything unusual. They hung canvas to catch the rain. Steamers raised sails to save coal. Frigate birds wheeled in the cantaloupe dawn.

Galveston spun through space at nine hundred miles an hour. The trade winds blew. Great masses of air shifted without a sound.

Somewhere, a butterfly opened its wings.

Violent Commotions

DESPITE THE GREAT demands of a nineteenth-century farming life, Isaac and his brother, Joseph, remembered the world of their childhood, in the knob-hilled terrain of Monroe County, Tennessee, as an Eden-like realm through which they wandered with little parental restraint. As a hobby, and to raise spending money, Isaac trapped muskrat, mink, and otter. He rose early to check his traplines before his daily chores began. His chores began at 4:00 A.M. He was six years old.

The Cline farm was among the richest in the knobs. In fall, at acorn time, passenger pigeons gathered in the oak trees in such great numbers they hid the treetops. The land was lush with apples, peaches, strawberries, and persimmons. Ghosts populated the black places under its forests. Isaac's uncle swore as fact that once during a hunting trip he had seen a headless woman who told him she was searching for a jug of whiskey buried fifteen years earlier by her husband. Stories circulated of a strange apelike creature spotted in the hills, and these too seemed like country tales, until the day armed officers captured a naked "wild man" and penned him at the center of town. Sinkholes could open overnight. One swallowed Joseph's plow. Another turned Boyd's Pond, a swimming hole on the Cline farm, into what Joseph called "our most thrilling devil's haunt"—the place where a boy was said to have boasted he would "swim the pond four times or go to hell." The boy finished the fourth circuit when the water began to whirl around him. He struggled, threw his arms up in panic, and plunged from view.

The law of convenient epiphany would place the trigger for Isaac's decision to become a meteorologist in the funnel of a tornado that swept into nearby Fork Creek Valley one Saturday night, lifted the bed of a sleeping child, and deposited the bed in an orchard one hundred yards away, the child still aboard and safe, the bed intact. Or perhaps in the great skeletons of lightning that clutched the sky so many August nights. These things played a part, no doubt. Lightning was barely understood, tornadoes not at all. To a boy in a land of ghosts and wild men, how could they not be alluring?

But other forces played on Isaac. He came of age in a time of broad technological awakening, in an America transformed by steam and telegraphic communication. He read everything by Jules Verne. Between bouts of plowing, while giving his mule, Jim, a rest, he would join Phileas Fogg and Captain Nemo on their elaborate adventures. Isaac loved science—his greatest dream was to write a scientific treatise on something, anything, as long as it resonated the world over—but he also loved the Bible, so much so that toward the end of his years in high school his friends urged him to become a preacher. At sixteen, he entered Hiwassee College in Tennessee, where he studied mathematics, physics, chemistry, Latin, and Greek. A few friends had set their sights on becoming lawyers, and for a time Isaac joined them in reading the works of Sir William Blackstone, the eighteenth-century English jurist, but never with a serious desire to practice. "I first studied to be a preacher, but decided that I was too prone to tell big stories," he later explained. "Then I studied Blackstone for a while and soon learned that I was not adept enough at prevarication to make a successful lawyer. I then made up my mind that I would seek some field where I could tell big stories and tell the truth."

He chose the weather.

ACTUALLY, THE WEATHER chose him.

Gen. William B. Hazen, in charge of the U.S. Signal Corps since 1880, wanted only the best men for his new weather service. Smart men, moral men, scientific men, but above all, strong men capable of wading against a mounting sea of skepticism about the corps' ability to report and forecast the weather. He wrote to college presidents asking them to recommend likely candidates from their graduating classes.

The president of Hiwassee College, J. H. Bruner, recommended Isaac.

"I accepted with pleasure," Isaac wrote, "for it was just the kind of work I wanted." General Hazen telegraphed instructions directing him to report to Washington on July 7, 1882.

ISAAC REACHED WASHINGTON'S Pennsylvania Railroad Station early on the morning of July 6. He was twenty years old and had spent his entire life in the hollows of Tennessee, but suddenly his world got much larger. Gigantic. The minute he stepped from the train he found himself standing where a president's blood had flowed. A marker showed the exact place where President James Garfield had been shot one year earlier by Charles J. Guiteau. Guiteau was hanged the week before Isaac's arrival. Now the platform was crowded with men whose great bellies and muttonchop whiskers spoke of power. Already the air was sticky and hot. It smelled of horses and smoke. The men wore black suits. They did not appear to suffer in the heat, but the air carried a certain added pungency. Never had Isaac seen so many people gathered in one place, amid so much noise and such a rich battery of scents. The whistles of locomotives shrieked; their boilers hissed. He heard an intermittent ringing and knew instantly it came from telephones somewhere within the station. Shiny black cabs clattered to the station doors, hailed by porters pushing high-wheeled handcars. Isaac saw telegraph poles so heavily strung with wire they looked like the backs of grand pianos. And there was talk of still more wire—that soon cities like New York, Philadelphia, and Washington would be lighted with electricity.

Isaac was exhausted, lonely, thrilled. He took a cab to the hotel booked by General Hazen, and there spent the rest of the day indulging in a very uncharacteristic pursuit: doing nothing. Partly it was the fatigue. But mainly this young man who had trapped the night forests of Tennessee at the age of six was frightened. He had never been in a city this big before. He was afraid even to let the hotel out of his sight.

He might have been a lot more anxious if he had known of the controversy that swirled at that moment around the Signal Corps, and of the scandal that triggered it, a scandal whose shock waves would roll forward like a storm swell to shape the events of Saturday, September 8, 1900.

But that night the only thing swirling seemed to be the mosquitoes clouding the gas lamps on the street below.

THE CRIME ITSELF could have happened in any bureau of the government, the juxtaposition of money and men always a chancy thing. That it happened within the Signal Corps, however, gave it an incendiary power beyond expectation. It had the effect of undamming a reservoir of complaint.

The corps had grown accustomed to controversy ever since Congress designated it the mother agency for the nation's first weather service. "Meteorology has ever been an apple of contention," wrote Joseph Henry, the first director of the Smithsonian, "as if the violent commotions of the atmosphere induced a sympathetic effect on the minds of those who have attempted to study them." Some critics argued men *should* not try to predict the weather, because it was God's province; others that men *could* not predict the weather, because men were incompetent. Mark Twain, merciless as always, parodied the government's efforts: "Probable northeast to southwest winds, varying to the southward and westward and eastward, and points between, high and low barometer swapping around from place to place, probably areas of rain, snow, hail, and drought, succeeded or preceded by earthquakes, with thunder and lightning."

But this new controversy was different. In 1881, police arrested Capt. Henry W. Howgate, chief financial manager of the Signal Corps, for embezzling nearly a quarter million dollars, this in an age when dinner at a nice restaurant cost thirty-five cents. He was arrested, convicted, and jailed. In the spring of 1882 prison authorities allowed Howgate to go home under guard to see his daughter, who was then visiting from Vassar. He escaped and was still at large when Isaac arrived in Washington.

For the weather service's critics, the Howgate scandal was the last straw. Secretary of War Robert T. Lincoln launched an investigation of the Signal Corps with particular emphasis on the service. He found it had few financial controls, a very limited pool of experienced forecasters, and a training academy—Fort Myer—that spent a lot more time putting men through cavalry drills than teaching them to forecast the weather. General of the Army P. H. Sheridan, around whom the aura of Civil War heroism still glowed bright, declared Fort Myer a waste of money. The all-important Chicago Board of Trade filed a formal petition with Congress demanding reform. Complaints also rose from within the Signal Corps itself, where some veteran military officers, among them a Major H. H. C. Dunwoody, opposed a push by General Hazen to conduct primary research into the causes and character of weather. Dunwoody objected in particular to Hazen's hiring of civilian scientists like Cleveland Abbe, easily the most prominent practicing meteorologist of the nineteenth century. The assault got personal when a Pennsylvania congressman accused General Hazen of cowardice in the Battle of Shiloh.

There were many things you could be in the new America, but a coward was not one of them.

. . .

AN ARMY SURGEON examined Isaac. He saw a lean young man of middle height with angular features, lively dark eyes, and an expression of sobriety that made you want to tell him some awful joke just to see if he could laugh. The surgeon had seen many boys like this, but under very different circumstances, and he wanted to tell this boy not to be so frightened, that his next stop was Fort Myer, not Bloody Run. Like most boys from the country, Isaac's face was sun-torched to a point about three-quarters up his forehead where his skin turned trout-belly pale. The boy had good hands. Strong, weathered, nicked. Enterprising hands. The doctor pronounced him fit.

Isaac and three other new men climbed into a wagon led by two strong horses and driven by a man in uniform. The wagon took them west through a neighborhood the driver called Georgetown, where three- and four-story brick houses stood jammed side by side. The wagon turned south and clattered across the Georgetown Bridge into Virginia, where it continued to climb until it reached Arlington Heights.

Even in the steam of that hot afternoon, the view was stunning. To the east was the great dome of the Capitol gleaming in the heat. A mile or so closer was the Willard Hotel and the tuft of forest that masked the president's mansion. A great stone tower dominated the landscape. It rose hundreds of feet into the sky and dwarfed every other building in sight. The tower was not yet finished. But how much higher could it possibly go? Nearer at hand, Isaac caught flashes of the Arlington mansion of Robert E. Lee and the great cemetery now aborning on its grounds.

The first soldier to greet Isaac was 1st Sgt. Mike Mahaney, a gruff Civil War veteran who showed Isaac to an oblong room with one window, running water, two double desks, and four beds. The fort's commander, Capt. Dick Strong, his natural seriousness amplified by his heavy beard, welcomed the new men and gave them his stock charge: "You will cheerfully obey all orders without question and refrain from saying anything either commendatory or condemnatory."

Isaac received a cavalry saber as part of his official kit. He loved its heft, and its cold hard lines, and how it evoked the stories he had heard men tell of Pickett's Charge. Soon Isaac found himself on horseback, learning how to kill men at a gallop—even though American military strategists, horrified by the carnage of the Civil War, had by then lost their taste for cavalry assaults. Isaac was a fine backcountry horseman, and caught on so quickly that Sergeant Mahaney placed him in charge of a squad of other recruits, some of whom had come from big cities and had never ridden horses.

Isaac led them around the track at vicious speeds, forcing some to wrap their arms around their horses' necks.

This could not have won him many friends.

THE HEART OF the weather service, and the thing that had to exist before there could even be such a service, was the telegraph. It allowed for the first time in history the rapid, simultaneous transmission of weather observations from stations thousands of miles away.

At Fort Myer, Isaac took apart and rebuilt telegraph transmitters to learn what caused the "click." A badly mauled telegraph pole stood in a squad room where its top extended into the skylight. Isaac learned to climb the pole and to string telegraph wire.

He also learned to send and receive messages and to use a special code developed by the weather service to save time and reduce the costs of transmission. The word *madman* indicated a morning barometric pressure of 28.33 inches. A wind of 57 miles an hour was *embalm*. The code word for a wind of 150 miles an hour was *extreme*. The cipher allowed a telegraph operator to pack a lot of information into just a few words. One example: "Paul diction sunk Johnson imbue hersal." Decoded, it meant: "St. Paul, 29.26 inches barometric pressure, −4 degrees temperature, wind six miles per hour, maximum temperature

10 degrees, dewpoint −18 degrees. This observation was at 8:00 P.M. and the local prediction called for fair weather."

But the service insisted that its men also know the tried-and-true visual methods of military communication. Isaac learned how to send messages using flags, torches, and the heliograph, which used a mirror to send bursts of light over long distances and was deployed later, in April 1886, during the Army campaign to capture Geronimo. Signal practice was awkward and difficult, especially at night when it required torches. These nocturnal sessions frequently involved "midnight travel in the rain, over muddy roads in black darkness, the horses choosing the proper route, as we could not," recalled H. C. Frankenfield, who also arrived at Fort Myer in 1882. Two decades later the bureau would assign Frankenfield the task of figuring out where the great hurricane of 1900 had come from.

Isaac became adept at signaling in every medium, but most recruits did not take this aspect of their training very seriously. They did not take much of anything seriously. Often recruits told each other in advance what messages they would send. One lieutenant deliberately marched a squad of new recruits double-time off the edge of a three-foot-high porch. Another officer, seeking to impress a carriage full of young women, suddenly ordered his squad to signal the word *asafoetida,* a medicinal ingredient that few knew how to spell. This prompted a moment of stunned silence, followed by a great flapping of flags evocative less of an elite signal squad than a flock of startled pigeons.

One morning a recruit named Harrison McP. Baldwin, the clown of his class, raced out in the predawn light for morning rifle drill, and executed without flaw all the required maneuvers.

Without his rifle.

No one noticed.

Years later, Baldwin went to work for Isaac Cline in Galveston. He was an able clown, an abysmal weatherman. It was a failing that Isaac would find intolerable, but one that probably saved Baldwin's life.

THE STORM

Monday, August 27, 1900:
15.3 N, 44.7 W

IT ADVANCED SLOWLY. Eight miles an hour, maybe ten. It moved west and slightly north and covered about two hundred miles a day, roiling the seas and erecting an electric wall of clouds visible to ships far outside its arc of influence. The first formal sighting occurred Monday, August 27. The captain of a ship at latitude 19 N, longitude 48 W, in the open sea below the Tropic of Cancer halfway between Cape Verde and the Antilles, noted in his log signs of unsettled weather. He recorded winds blowing from the east-northeast at Force 4, a "moderate breeze." Thirteen to eighteen miles an hour. His barometer showed 30.3 inches.

He dismissed the storm as a distant squall.

What Isaac Knew

BETWEEN BOUTS OF mounted swordplay, Isaac journeyed deep into the mysteries of weather. Meteorology was an emerging science rooted not so much in rigorous research as in stories and adventures, which only enhanced the mystery. By gaslight, with the bells of Washington tolling softly in the summer steam, he immersed himself in the millennial quest to understand wind, and in the hunt for the Law of Storms, one of the driving scientific explorations of the nineteenth century. He found it all as compelling as anything by Verne, a great sweeping saga full of crimson clouds, hundred-foot waves, and strange occurrences. He read how men caught in the fiercest storms found the decks of their ships carpeted with exhausted horseflies and how the survivors of a colonial hurricane emerged to find deer stranded in trees. In the Caribbean, wind had lifted cannon.

Weather was a national obsession and had been for centuries. Countless men, including some of the most prominent of their times, kept daily track of the weather and often for decades on end. Thomas Jefferson kept a lifelong weather journal and on July 4, 1776, despite certain other pressing matters, noted the temperature in Philadelphia to be a lovely 76 degrees. Samuel Rodman Jr., a prominent Massachusetts merchant, and his son Thomas together produced an uninterrupted daily record that began in 1812 and continued for three-quarters of a century. Such detailed journals told nothing about the fundamental forces that powered the weather, but they gave the men who kept them a sense of mastery over nature. By recording the weather, quantifying it, comparing it year to year, they demystified it at least to the point where storms ceased to be punishments meted by God.

But with God at least partly out of the way, the mystery only deepened. The first "scientific" definition of wind, by Anaximander, a Greek natural philosopher, would have seemed laughably primitive to Isaac, but for its time six centuries before the birth of Christ, it was a wonder of ingenuity. He called it "a flowing of air."

But what was air?

The first person to show conclusively that air had substance was Philo of Byzantium during the third century B.C. He attached a tube to a glass globe, then inserted the open end of the tube into a dish of water. When he placed the globe in shadow, the water rose within the tube. When he exposed the globe to sunlight, the level fell. "The same effect," he wrote, "is produced if one heats the globe with fire."

He did not know it, but he had stumbled upon the fundamental engine that drove the world's weather and that two thousand years later would power the ships of Columbus and his peers briskly over and with dismaying regularity *under* the seas. He had missed the broader question: If heat could cause a small volume of air to drive water up and down a tube, what could it do to the vast sea of air that covered the world?

Aristotle proved beyond doubt that air had mass when he demonstrated that a container filled with air could not also be filled with water. Did this mean that air had weight?

Aristotle flattened an airtight leather bag and weighed it, then filled the bag with air and weighed it again. Nothing changed. He concluded, erroneously, that air was weightless.

The world tumbled forward. Over the next fifteen centuries, the definition of wind did not advance very far beyond Anaximander's "flowing of air." In A.D. 1120, before Europe rediscovered the great works of the Greeks, Adelard of Bath, an English monk, thought he had stumbled upon something new.

With the sobriety of a man humbled by his own genius, he wrote: "I think that wind is a species of air."

As MEN VENTURED beyond the bounds of their accustomed territory, goaded by riches and glory, they encountered strange new meteorological phenomena. Early mariners discovered the miraculous trade winds that blew their ships toward the Indies. But they also discovered the doldrums at the equator and, just north of the trades, another realm of stillness that they named the Horse Latitudes, where half-dead crews becalmed for weeks cast their horses overboard to conserve drinking water.

Early captains learned also that these new seas harbored the exact opposite of doldrums, monster storms with cunning lulls during which the sun would shine and the winds cease, seducing unwary crews into believing the worst was over. Isaac learned that the first European to encounter such storms was the ever-charmed Columbus, and how the weather of the Indies revealed itself to him gradually, as if to prepare him for his first true hurricane. That storm occurred during his fourth and final voyage with such discriminating ferocity it sparked accusations that he had conjured it through magic—a not-unreasonable charge given the mysticism of the age, and the storm's result.

COLUMBUS SET OFF on his first voyage on August 3, 1492, from Palos, Spain, with a fleet of three tiny caravels, the *Niña,* the *Pinta,* and the *Santa María.* By nineteenth-century standards, the three vessels hardly qualified as ships. They were large boats crewed skimpily with a few experienced sailors and adventure-hungry boys. Not only did Columbus and his captains have no means of determining the exact location of their ships in the featureless blue of the ocean, they also carried none of the meteorological tools that mariners in Isaac's time took for granted.

After overcoming a few technical problems, the ships caught the trades and made quick, untroubled progress. The weather was perfect:

clear blue skies, brisk and steady winds shoving big cotton clouds over the horizon, cool nights and balmy days, the overall effect one of languid, sloe-eyed sensuality. "The weather was like April in Andalusia," Columbus wrote, "the only thing wanting was to hear nightingales."

But something curious did occur during that first voyage. A lookout saw them first, rising a long way off. Astonished, he sounded the alarm.

IT WAS SEPTEMBER 23, the fleet's exact position unclear but the weather good, skies bright, no sign of a storm on any horizon. Nonetheless, the lookouts spotted immense swells marching slowly and silently toward the ships. Columbus and his captains turned the fleet into the oncoming seas and watched open-jawed as the surface of the ocean rose in great oil-smooth hills of blue and green. The swells lifted the ships to exhilarating heights but posed no danger.

What Columbus did not know was that these swells were most likely the advance guard of a hurricane rising hundreds of miles away, well out of sight—the same brand of swell Isaac observed as he stood on the seat of his sulky in Galveston four centuries later.

The ships continued their journey; Columbus opened the gates to the New World.

The more time Columbus spent in the waters of the Indies, however, the more he saw the flaws in his original appraisal of Caribbean weather. Water spouts danced among his ships. Tropical rains fell as if from a ruptured cask. Squalls tore the sails from his spars. By the time of his final voyage, Columbus had learned that the seas of the New World were both seductive and deadly, but in the process had become adept at reading the tropical skies for signs of trouble.

He was ready for his first true hurricane.

FOUR YEARS BEFORE the storm, Ferdinand and Isabella, intending to reward Columbus, appointed him viceroy of the Indies. He reached

Hispaniola in August 1498 expecting to savor the perquisites of rank, but found rebellion and turmoil. When word came back to Spain that chaos, not the sovereigns, reigned in Hispaniola, Ferdinand and Isabella dispatched an emissary, Francisco de Bobadilla, to straighten things out. Secretly they had granted him extraordinary powers, which he demonstrated immediately upon his arrival. It did not help that as Bobadilla sailed into Santo Domingo harbor he saw seven Spanish corpses dangling from the gallows. Swaying palms were one thing; swaying countrymen quite another. He used the hangings as a pretext to arrest Columbus and lock him in chains, a degree of public humiliation that speaks clearly of some deeper passion filling Bobadilla's portfolio. Greed perhaps, but certainly envy.

In October 1500 Bobadilla marched the iron-laced Columbus through town and on board a ship, *La Gorda,* bound for Spain. Bobadilla himself took over the administration of Hispaniola. After returning to Spain, Columbus remained in chains for six more weeks before the sovereigns released him. He pleaded for the license and funds to conduct one more great voyage. In a sign of new warmth toward the admiral, Ferdinand and Isabella commanded Bobadilla to assemble all proceeds from trade and the mining of gold that were owed Columbus, and to place these in the custody of his designated agent. On March 14, 1502, the sovereigns granted Columbus another voyage. Like wise parents seeking to head off the wars of jealous children, they forbade him to stop at Hispaniola.

Columbus, delighted to be sailing again, set out with four caravels, and on June 29, 1502, found himself and his fleet off Hispaniola. He saw that a great convoy of thirty ships was being readied in the Ozama River at Santo Domingo for imminent departure, but did not know at the time that this fleet was carrying Bobadilla and a vast fortune in gold, including his own share. That Bobadilla had consigned Columbus's gold to the smallest and flimsiest of the convoy ships, the *Aguja,* was yet another mark of whatever hidden passion fueled his hatred. If any ship was likely to sink, it would be the puny *Aguja*.

Columbus had at least three good, practical, defensible reasons for what he did next: First, the departing convoy presented an excellent opportunity for getting mail from his own little fleet promptly back to Spain. Second, he wanted to trade one of his ships, a poor performer, for something a bit more spry. Third, the weather had taken an ominous turn, exhibiting the usual troika of storm signs: oily swells, oppressive heat, a red sky.

For all these good, practical, and defensible reasons, Columbus sent one of his captains ashore with a request to permit his fleet to enter the harbor, a clear violation of the sovereigns' orders.

The new governor, Don Nicolas de Ovando, only laughed.

Stung, Columbus led his ships to the leeward side of Hispaniola to place the mass of the island between the ships and the rising storm. He instructed his captains that if they became separated by the storm to meet in a harbor on Ocoa Bay, near what later became Puerto Viejo de Azua.

Meanwhile, with great fanfare—trumpets blaring, cannon roaring, banners streaming—the thirty-ship convoy ferrying Bobadilla and Columbus's gold sailed from Ozama and made for the Mona Passage, the strait between Hispaniola and Puerto Rico that connects the Caribbean to the Atlantic.

The storm was a full-fledged hurricane. Columbus's fleet, sheltered in the lee of Hispaniola, caught a glancing blow that nonetheless topped anything in severity that Columbus had so far confronted. "The storm was terrible," he wrote, "and on that night the ships were parted from me. Each one of them was reduced to an extremity expecting nothing save death; each one of them was certain the others were lost."

In a maneuver that went against customary marine practice, Columbus did not strike for open sea but instead brought his ship closer to shore to leverage further the windbreak afforded by the mountains of

Hispaniola. His ship survived. On Sunday, July 3, he sailed his caravel into Ocoa Bay, the designated meeting place. He saw no sign of the others.

As his ship rocked gently in the gorgeous blue, its decks quiet but for the sounds of repair, Columbus watched the entrance to the bay through thermals of humid air.

A lookout would have spotted it first as a glint of white against the settling sea. He cried out, then perhaps wished he had not, as the glint disappeared and the ship eased back into the turquoise quiet.

But another spark followed, a true sign now. Sails and finally a ship. Followed by another. And, impossibly, yet another.

All safe.

And what of Bobadilla?

The hurricane caught the convoy in the Mona Passage head-on, the eye passing close, perhaps directly overhead. It drove twenty of the gold ships to the bottom with all hands. One of these carried Bobadilla. In all, five hundred mariners lost their lives. A few ships, gravely wounded, fought their way back to Santo Domingo.

Only one ship of the original thirty made it to Spain: the puny little *Aguja,* carrying Columbus's gold.

THE ENIGMA OF air continued to command the attention of the world's greatest minds. In 1638, Galileo tried a variation of Aristotle's leather-bag experiment. He constructed an apparatus consisting of a glass bulb with an airtight valve. He weighed the bulb. Next he forced air into the bulb until it contained much more than its normal volume. Now when he weighed it he found a measurable difference.

So air *did* have weight.

In Galileo's time this was astonishing news. Air was invisible, yet it had weight. It was everywhere, piled high over the world. Therefore it must exert a force on every man, rock, and tree. The meteorological significance escaped Galileo, but five years later his discovery led to a famous series of experiments by Evangelista Torricelli, an Italian physicist who opened the single most important window into the forces that drive the world's weather.

He too began with a glass bulb, but attached to it a tube some "two cubits" long, a cubit being a vague unit of measurement equivalent to the distance between a man's elbow and the tip of his middle finger. He filled this tube with mercury, inserted the tube into a bowl also containing mercury, then watched the mercury in the tube fall until it stabilized about halfway between the bulb and the dish.

It never completely stabilized, however. Torricelli observed that it crept up and down at different points during the day and under differing atmospheric conditions. He did not come to this easily. Before he settled on mercury, he tried water. To get any observable effect, he had to use a glass tube sixty feet long, not exactly a device likely to win favor among mariners headed for Shakespeare's "vexed Bermoothes."

The term *barometer* arrived a decade or so later when Robert Boyle coined the name to describe his own air-weighing device, an instrument that so delighted the Royal Society, it resolved in 1668 to have a collection of Boyle's barometers built and dispatched to the far limits of the world. The proposal was never enacted, but by Isaac's time the barometer had become so well accepted as a meteorological tool that it wound up in all those places anyway.

Storm accounts got more and more detailed, kindling the imaginations of countless landlocked boys and providing the first scientific insight into the unique character of hurricanes. One of the most compelling writers of the seventeenth century was William Dampier, an Englishman who split his time between adventuring with buccaneers and patiently recording the natural phenomena he encountered on his far-flung voyages. Isaac considered him one of the great pioneers of meteorology. It was Dampier who gave the world its first detailed description of the lurid atmospheric colors that preceded such storms—the "brick-dust sky" that Isaac looked for but did not find as he scanned the Gulf horizon.

In 1703 a storm of great power and endurance brought the realities of cyclones to the heart of London itself. In giving England the worst storm of her history, it also advanced the literary career of Daniel Defoe, a forty-three-year-old editor and journalist with a taste for disaster. He knew a good thing when he saw it.

FOR TWO WEEKS in November 1703, a pod of strong gales paralyzed shipping off the coast of England. Outbound ships had to remain in port; inbound ships had to stay at sea. On Wednesday, November 24, the winds abated; by Thursday, hundreds of ships, including a contingent of Russian warships under ceremonial escort by the British man-of-war *Reserve,* began to move in a slow and graceful waltz over the rough "old seas" left behind by the storms.

The *Reserve* put in off Yarmouth. Her captain, convinced the worst was over, went ashore with his ship's surgeon and clerk to buy provisions. In Deal, a small town overlooking the treacherous Goodwin Sands near Dover, Mayor Thomas Powell spent the day at his full-time job as "slopseller," peddling supplies for seamen. In Plymouth, Henry Winstanley and a crew of workmen set out from the Barbican Steps on a fourteen-mile sail to Winstanley's controversial Eddystone Light to repair its failed beacon. His critics had charged the lighthouse was unsafe, to which Winstanley responded that his one wish was to be inside the structure during "the greatest storm that ever blew under the face of heaven"—one of those moments in history that begged for a burst of ominous music.

BY NOW BAROMETERS could be found not just in the possession of mariners and scientists, but also in some private homes. Scientists understood too that foul weather tended to be accompanied by falling barometric pressure, although why this should be the case remained a mystery. Late on Friday, November 26, the barometer owners of England saw the level of mercury begin to fall, then plummet.

The storm struck with such ferocity that Queen Anne was escorted into the basement of the Palace of St. James and there deposited in a wine cellar. Wind stripped the roof off Westminster Abbey and demolished over four hundred windmills, in some cases turning their mill sails so fast that friction set the buildings on fire. The wind hurled roof tiles like cannon shot.

The storm destroyed seven hundred vessels on the Thames within London, jumbling them into great piles of debris, bowsprits impaling stern cabins. A tangle of rigging and tackle lay over all as if a giant spiderweb had settled upon the wreckage. Along the Severn River, storm waters breached seawalls and drowned fifteen thousand sheep. Salt spray turned leaves white. Antonie van Leeuwenhoek, the naturalist, wrote how at eight the next morning, "I cast my eye upon my barometer, and observ'd, that I had never seen the quick-silver so low."

On land, only 128 people died, many killed by the collapse of fireplace chimneys.

At sea the story was different. If not for the clamor of wind and surf, what one would have heard that night up and down the coast of England was the thin cry of doomed men, stranded or adrift, many hanging from the tops of masts that now protruded only a few feet from the sea.

Off Plymouth something happened that most men would have dismissed as impossible. If one could count on anything in Defoe's time, as in Isaac's, it was a lighthouse.

UNTIL SHORTLY AFTER midnight, Friday night, residents along the distant mainland saw the reassuring beaconflash of the Eddystone Light. It proved that Henry Winstanley had succeeded in repairing the lamp despite the hurricane that must have welled up even as the work got under way.

After midnight, the light ceased to shine. When rescuers at last reached the lighthouse, or rather, the rock on which it had been built, they found nothing. The storm had scoured the light from the face of the earth. Only the barest trace of timber and masonry marked that anything at all had stood there, let alone a lighthouse.

Farther along the coast, several ships ran aground on the Goodwin Sands. Survivors hung in the upper masts and rigging of their ships until the tide receded, then climbed down to the now-exposed sands to await rescue, certain that the town they saw through the spindrift soon would send help.

The residents of Deal were aware of the sailors' plight. Some watched the stranded men through telescopes. "It must have been a sad spectacle," Defoe wrote, "to behold the poor seamen walking to and fro upon the sands, to view their postures and the signals they made for help, which, by the assistance of glasses, was easily seen from the shore."

Boats did set out from Deal, but not for rescue. Their occupants ignored the doomed men and instead probed the floating debris for valuable salvage. The men on the sands were fathers, husbands, lovers, and sons, "but nobody concerned themselves for the lives of those miserable creatures."

When Mayor Powell learned of his town's behavior, he was appalled. He pleaded with the local customs house to deploy its boats for rescue, but the official in charge refused. Powell tried to raise his own corps of rescuers, offering five shillings for every sailor saved. With the help of a few volunteers Powell seized the customs boat and by his example convinced some of the salvage crews to help. The rescuers saved two hundred men but could not return in time to save the many others still stranded when the tide returned.

In all, the great English cyclone of 1703 killed over eight thousand seamen aboard hundreds of ships. One victim was the man-of-war *Reserve*. As the storm intensified, her captain, surgeon, and clerk raced back to the wharf in Yarmouth, where all they could do was stand and watch as the seas consumed the ship and all aboard.

Men understood the hazards of hurricanes, but the fundamental engines that drove such weather continued to elude them. Where did wind come from? And what gave it such power?

By the early eighteenth century important pieces of the puzzle were in place. Air pressure could be measured, even at sea. Temperature scales at last allowed precise comparisons of hot and cold.

The most important piece, however, lay unrecognized, even though the underlying principle had been proven long before.

IN 1627, A very brave if melodramatic German mathematician, Joseph Furtenbach, aimed a loaded cannon into the sky in preparation for an experiment he hoped would provide the first real-world test of another of Galileo's theories: that the earth rotated on a fixed axis. This was high-wire science. If Galileo was right—and Furtenbach fervently hoped he was—a cannon ball fired straight into the sky would fall back to earth somewhere to the west of the cannon, while the earth's rotation carried Furtenbach safely east. If Galileo was wrong, the ball would fall to earth exactly at the point where it rose from the cannon, and Furtenbach would be dead.

He fired the cannon. As the ball soared into the sky, he hurried to the muzzle and sat on it. Skeptics in the audience no doubt stepped back a respectful distance, wary not only of the descending ball but also of the likely splash of viscera. How the seconds must have dragged as that ball whined into its descent, the smile on Furtenbach's face growing fixed, the more squeamish members of the audience raising their hands to cover their eyes but peeking of course through the latticework of fingers. . . .

Thwump.

Silence.

Furtenbach slid from the muzzle, his head and smile intact. To the west—a small crater. Proof at last. The earth did spin.

It was Edmund Halley, of comet fame, who recognized that this rotation might have a powerful effect on the earth's weather. Seeking to explain the trade winds, Halley argued that the sun's rays fell most consistently upon the equator. As the sun moved over the earth, it caused successive parcels of air to rise. Other, cooler air rushed in to fill the space and followed the sun around the globe in a steady rush of wind.

A compelling theory, but it had a significant hole: It could not explain why the prevailing easterlies of the trade belt suddenly gave way north of the Horse Latitudes to winds blowing in exactly the opposite direction.

What Halley failed to take into account was the *shape* of the earth: the fact that the world moves more slowly in New York City—although no New Yorker would ever concede it—than in Key West. In 1735, George Hadley, often confused with Halley, crafted an explanation of the trades that was so compellingly simple it remained the accepted theory even through Isaac Cline's Saturday.

HADLEY RECOGNIZED THAT an object anchored near the north pole and another near the equator traveled through space at different speeds: Both objects, being attached to the same planet, had to complete one rotation within the same period of time, but the object at the equator had to cover a much greater distance and therefore had to move a lot faster. The air at each location, Hadley saw, also moved at these differing velocities.

He agreed with Halley that as the sun heated the equator, it caused air to rise, and cooler air flowed in to take its place. But Hadley proposed that the cool replacement air would retain its polar velocity. The farther south it went, the slower it would seem to travel relative to the ground below. Anyone encountering this slow-moving mass of air would experience it as a wind that veered to the right of its direction, or toward the west. These were the trade winds.

Conversely, Hadley saw, air migrating north would seem to *accelerate* relative to the ground. As it cooled, it would descend but retain its faster equatorial speed. Observers on the ground would perceive this as a wind blowing toward the east, or veering to the right of its northward course. This wind, Hadley argued, produced the steady breeze north of the Horse Latitudes that blew the explorers back home.

A century later, a French mathematician, Gaspard Coriolis, worked out the mathematics of all this, to prove that any object moving over the northern hemisphere would seem to veer to the right, while any object moving over the southern hemisphere would appear to veer left. Isaac, in his 1891 talk to the Galveston YMCA, gave a cruelly detailed explanation of the Coriolis effect. The crowd listened with iron concentration. ". . . At latitude 30 degrees the velocity of the earth eastward is 897 miles per hour, and at 45 degrees it is 732 miles per hour, or 165 miles less. Now, if a mass of air in a quiescent state were transferred instantly from the thirtieth parallel to the forty-fifth parallel it would be found to have a relative motion eastward of 165 miles per hour greater than that of the parallel arrived at, and if it had been transferred from 45 degrees to 30 degrees, with the motion which it had at 45, it would be 165 miles slower than the earth at the thirtieth parallel, and this would give a relative velocity westward of 165 miles per hour."

A twentieth-century audience would have shot Isaac dead.

HADLEY'S THEORY DID little to advance man's immediate understanding of storms in general and hurricanes in particular. Meanwhile, the danger grew. Ship traffic increased. Nations deployed battle squadrons to protect their interests. No single period highlighted the threat to national defense more than the wild hurricane season of 1780, during which three intense hurricanes scoured the Caribbean in a period of two weeks and impartially ravaged the forces of France, Spain, and Britain, even as these nations harried one another in the war-convulsed seas of the Americas.

The first hurricane arrived October 3 and leveled the Jamaican town of Savanna-la-Mar, and in the process overtook scores of British warships. Hundreds of seamen simply vanished. "Who can attempt to describe the appearance of things upon deck?" wrote Lt. Benjamin Archer, who survived the foundering of the forty-four-gun *Phoenix.* "If I was to write forever, I could not give you an idea of it—a total darkness all above; the sea on fire, running as it were in Alps, or Peaks of Teneriffe; (mountains are too common an idea;) the wind roaring louder than thunder (absolutely no flight of imagination,) the whole made more terrible, if possible, by a very uncommon kind of blue lightning."

The second hurricane, called simply the Great Hurricane, struck Barbados on October 10 and 11, killing 4,326 people on that island alone. The toll throughout the Indies reached 22,000. Britain's Sir George Rodney, deeply shaken by the disaster, described what remained of Barbados: "The most beautiful island in the world has the appearance of a country laid waste by fire, and sword, and appears to the imagination more dreadful than it is possible for me to find words to express."

The storm lurched into French territory next and sank at least forty ships in a French convoy off Martinique, with a loss of five thousand soldiers.

The third hurricane struck just as Spain's Admiral Don Jose Solano was leading a force of six dozen ships and four thousand soldiers for a surprise attack against the British at Pensacola. The storm so damaged and dispersed the fleet, the admiral called off the attack. In keeping with the early custom of naming storms after prominent victims, the hurricane became known as Solano's Storm.

Together the three hurricanes did so much damage to Britain's Caribbean forces that the Admiralty canceled a secret plan to seize Puerto Rico from the Spanish.

No navy could have made such short work of the military might of the world's greatest powers. Clearly hurricanes posed a greater menace than any single nation's forces. But what could one do? Captains could not even measure the velocity of the winds they encountered, for no effective means existed of measuring wind from a rolling, heaving ship. Sir Francis Beaufort tried to solve that problem by devising a wind scale that allowed mariners to gauge the intensity of wind by the look of seas and sails. Force 0 meant winds so light a ship could not move. Force 12 was a hurricane, when no sail could be exposed. Beaufort's intent was to bring uniformity, and with it comparability, to weather observations made at sea. His scale included no actual wind velocities—these were added much later. The first captain to use the scale in an official log did so on December 22, 1831, the first day of a voyage of exploration. The captain was Robert Fitzroy; his ship's company included a naturalist named Darwin.

Hurricanes, once such a surprise to Columbus, became lodged firmly in the public psyche as just another hazard of venturing upon the sea— acts of God, *still,* and against which one could do nothing. With tragic regularity, captains sailed their ships right into the worst storms that ever danced upon the earth. Seamen resigned themselves to the inevitability of hurricanes and prayed they would never have to experience their full fury. But others were not so willing to surrender. They began an earnest search for the elusive "Law of Storms," the physical code that scientists hoped would help mariners predict and avoid—perhaps even profit from—the hurricanes and typhoons that so threatened the welfare of nations.

A hurricane set the hunt in motion.

ON SEPTEMBER 3, 1821, a hurricane moving up the coast from Cape Fear made landfall near New York City, and continued north well into New England. Soon after the storm a thirty-two-year-old saddler named

William Redfield, son of a long-dead sailor, took a trip on horseback through Connecticut and happened to notice something unusual in the landscape around him. Near Canaan, in northernmost Connecticut, the trees had fallen in a direction exactly opposite that of the toppled trees he had seen farther south.

After his return home, Redfield made a careful study of the hurricane. He collected fragments of detail about the storm from newspapers, letters, ships' logs, and other sources, and in the process became the first man to track a hurricane from first sighting to last. His interest expanded to include other hurricanes, which he pursued with equal zeal. His first paper, "On the Prevailing Storms of the Atlantic Coast," appeared in 1831 in the *American Journal of Science,* and quickly became a classic of meteorology. He concluded there could be only one explanation for the changing pattern of damage he had encountered: "This storm was exhibited in the form of a great whirlwind."

Redfield's meticulous research caught the attention of a British naval officer, Lt. Col. William Reid, who had been dispatched by King William IV to Barbados to supervise the reconstruction of British interests there in the wake of yet another disastrous hurricane, this the great "Barbados-to-Louisiana Hurricane" of 1831, which killed over fifteen hundred people. Reid too became obsessed with hurricanes. After his return to England, he adopted Redfield's tracking techniques and in turn ignited the storm-watching passions of a countryman, Henry Piddington, who applied the same techniques to the unfathomably deadly storms of the Bay of Bengal. It was Piddington who coined the word *cyclone,* from the Greek for "coils of a snake," and it was his research that resonated most darkly within Isaac Cline on Saturday, September 8, 1900.

Piddington reconstructed a cyclone that struck the bay town of Coringa in December 1789. "The unfortunate inhabitants of Coringa saw with terror three monstrous waves coming in from the sea, and following each other at short distances. The first, sweeping everything in its passage, brought several feet of water into the town. The second augmented these ravages by inundating all the low country, and the third overwhelmed everything." The three waves killed at least twenty thousand people, although the final toll was beyond tally. "The sea in retiring left heaps of sand and mud, which rendered all search for the property or bodies impossible."

Isaac read Piddington's work. It would come back to him years later on the beach at Galveston. "I had studied the meagre information available relative to tropical cyclones," Isaac wrote. "I had read of the Calcutta cyclone, October 5, 1864, which caused a storm tide 16 feet deep over the delta of the Ganges and drowned 40,000 persons, and the Backergunge cyclone of October 31, 1876, which caused an unprecedented storm tide ranging in depth from 10 feet to nearly 50 feet over the eastern edge of the delta of the Ganges, and drowned at the lowest estimates 100,000 persons." At that point, however, he was only thinking in terms of waves. He still had no appreciation of how similar the undersea landscape, or bathymetry, of Galveston Bay was to that of the Bay of Bengal. That would come later.

Piddington, in his immensely popular text *The Sailor's Horn-Book for the Law of Storms,* offered practical advice for coping with hurricanes. He included transparent storm cards, or "horn cards," which showed the direction of wind at various points in a cyclonic circle. A mariner could match the winds he was experiencing with the winds marked on the card and thus determine where in the body of the storm his ship was located and thus how to avoid sailing toward what Piddington called "the fatal centre." With these storm cards, Piddington wrote, "*you have the hurricane in your hand.*"

It all sounded good and precise on paper, but hurricanes still came by surprise, still killed by the thousands. As one nineteenth-century captain put it, "if the centre *always* bore eight points from the direction of the wind; if the wind *gradually* increased in force as we near the centre;

if the wind veered *gradually* in all parts of the storm; and if the centre were the *only* dangerous part of it, then the avoiding of a hurricane would be very simple."

WHAT ISAAC DID not learn much about at Fort Myer was forecasting, a black and dangerous art that only a few men in Washington were allowed to practice. Incorrect forecasts eroded the faith of a public already skeptical of the service's prowess and worth. A few newspapers had taken to running the service's weather forecasts opposite the often-superior forecasts of astrologers and assorted weather prophets. To help ensure that the best men got deployed to the field, the weather service gave its Fort Myer trainees a rigorous examination. The top scorers won immediate assignment as assistant observers to posts throughout the country.

At one point the test asked each trainee to choose a scientific mission related to meteorology that each could pursue while conducting the routine work required in a weather station. The chief did not want his observers just sitting around between weather observations, a wise policy, given the sex scandals, grave robbing, and other incidents that would soon surface and further undermine the weather service's reputation. Isaac gave a beauty queen's answer—that he wanted to do something that would "give results beneficial to mankind."

Isaac scored in sixteenth position, and the service promptly assigned him to Little Rock, Arkansas. When not recording temperature and barometric pressure, he was to investigate how climate shaped the behavior of Rocky Mountain locusts, said to be swarming the countryside. To Isaac, this was the fulfillment of a dream. "I was twenty-one years old," he wrote, "the world was before me and my enthusiasm was such that I thought I could do any thing that it was possible for man to accomplish."

THE STORM

Tuesday, August 28, 1900:

16 N, 49.3 W

THE VORTEX GAINED definition. Rivers of air flowed toward its center. The earth's rotation drove them to the right, but each right-veering gust imparted to the vortex a left-hand spin, just as a glancing blow on the right side of a cue ball will cause it to spin left. The arriving winds lowered pressure. As the pressure fell, air moving toward the storm gained velocity. The stronger winds drew more water vapor from the sea, which fed the clouds around the center of the vortex—releasing more heat and driving the pressure still lower.

On Tuesday, August 28, the storm overtook a ship located about three hundred nautical miles southeast of Monday's first sighting. The ship's log noted winds from the south-southwest, the bottom right rim on a Piddington horn card. The wind was stronger, Force 6, twenty-five to thirty-one miles an hour.

Guy wires whistled.

Dirty Weather

IT WAS WINTER. Isaac's train passed through an austere landscape of grays and browns, the trees like upended spiders, but to him all of it was dazzling. "Something new, something of interest and beauty unfolded before my eyes all the time." He arrived in Little Rock just before the state legislature passed a bill that resolved a long-festering controversy. Henceforth, the new law declared, the legal pronunciation was "Arkansaw."

Isaac's boss assigned him responsibility for weather observations to be made at five in the morning and eleven at night. In between he was to put together bulletins for the station's customers and collect weather dispatches cabled each day by a network of railroad agents.

He did not find any locusts. "They evidently learned that I had been put on their trail and disappeared." But he did find another means of filling his time.

The University of Arkansas's medical school was only three blocks from the station. Medicine, Isaac reasoned, would provide not only a productive way to fill his day, but also satisfy the Signal Corps' requirement that its observers pursue a scientific endeavor related to their daily duties. He could study how weather and climate affected people, a new field and one that "could not evade me as the Rocky Mountain locusts had done." He enrolled in the middle of the 1882–83 school year, and found his work and study schedules complemented each other. "The one gave me a rest from the other," he wrote, "and I never became tired."

He graduated from medical school on March 29, 1885. Five days later General Hazen placed him in charge of a weather station at Fort Concho, Texas. The nearest town was San Angelo, whose residents described the place as hell on wheels. Hazen directed Isaac to travel by rail to Abilene, Texas, and there to catch a stagecoach for the one-hundred-mile journey to the fort. But when Isaac checked his Rand McNally Railroad Map he could not find Abilene.

It did exist, the railroad agent assured him. It was just too new to be on any map. A cattle boom had created the town overnight.

As the agent prepared Isaac's tickets, he told him a story, the first of many unsettling stories Isaac would hear about the West in the days before his departure.

The railroad had just reached Sweetwater, the agent explained, Sweetwater being another spanking-new town some thirty-five miles west of Abilene. Just a few days earlier half a dozen Chinese railroad workers had been gunned down by a group of drunken cowboys. The sheriff arrested the killers and brought them before Sweetwater's brand-new judge, who had also opened a saloon.

The judge considered the case, pursed his lips, opened a couple of law books just to make sure his first bone-deep feelings about the case were correct, then issued his judgment: "Gentlemen," he ruled, "I have examined the laws of the United States carefully and I do not find any law which says that a white man shall be punished for killing a Chinaman."

The judge, named Roy Bean, let the killers go.

Isaac paid close attention to one fragment of advice. "I was told that well-dressed men often had their hats shot off their heads and their good clothes pulled from their backs."

In Little Rock, Isaac had become a dandy. He had adopted, whole-heartedly, the fashion then in vogue among the city's doctors. On his rounds at Little Rock's Charity Hospital he wore a Prince Albert brown beaver suit, silk top hat, and kid gloves. And carried a cane.

He was twenty-three years old.

He was as good as dead.

When Isaac climbed aboard his westbound train, he wore a battered old suit from his last days in Tennessee. He could not bear to leave his fancy clothes behind, however. He hid them under the false floor of his trunk.

ISAAC ARRIVED IN Abilene under a gunmetal sky, the city awash in mud and scented with horse manure and fresh-sawed lumber. He heard the torn-fabric scree of ripsaws and the sound of hammering as joists and beams went up in new buildings around town. Cowboys strolled around in high boots and spurs the size of daffodils, and wore pistols shoved into their waistbands. He had entered a territory as alien to him as anything he could have concocted in a daydream. Here before him was the West of Jules Verne's *Around the World in Eighty Days,* in which Phileas Fogg, an Isaac-like character of precision and rigor, raced across the Great Plains during the American leg of his journey around the globe.

Isaac learned that the coach to San Angelo would not arrive until the next morning. He tried Abilene's one hotel, but found it full. A railroad agent told him about a room for rent over a saloon.

At the entrance, Isaac encountered a porter mopping the wooden sidewalk. The water had a red tinge to it. Perhaps joking, Isaac said, "That looks like blood."

"Yes sir," the porter said casually, without breaking his rhythm. He explained that four cattlemen had gotten into a gun battle. These were not just ordinary cowboys, he said, but well-off ranchers with large herds. Now all four were dead.

Isaac stepped past. He checked in and climbed the stairs to his room. "My head," he wrote, "did not rest easy that night."

IN THE MORNING, things looked better. The sun was bright, the air cool and scented with bacon, coffee, and sawdust, the fragrance of a brand-new country. The landscape was amber, pierced by long black pickets of shadow. Isaac was twenty-three years old in a new country in a world where anything was possible. He was in the thick of it when everyone else back home could only read about it in the newspapers and in Jules Verne and in the thousands of dime novels about Buffalo Bill Cody. Isaac was a pioneer in a new science, a prairie Dampier, at a time when an ordinary man with patience and a knack for observation could change forever the way the world saw itself. Far to the north in the Bad Lands of the Dakota Territory another young man, Teddy Roosevelt of New York, was busy "pioneering" along with other East Coast blue bloods like Frederic Remington and Owen Wister, later to write *The Virginian,* who hoped to experience the frontier life before it disappeared. Roosevelt called this way of living "the pleasantest, healthiest, and most exciting kind of life an American could live."

The stage arrived clotted with mud, then set off again in a great jangle of energy, pulled by four horses and rocking on its springs like a bark in heavy swells. The coach was scheduled to cover the one hundred miles to San Angelo by late afternoon with a team change every thirty miles, but a rain-engorged stream halted the journey. The driver told Isaac and his fellow passengers they would have to spend the night alongside the creek until the next scheduled coach could arrive from the opposite direction. The driver would then ferry the group across the creek, using a boat kept at the crossing for just such emergencies. The fresh coach would return to San Angelo.

The sole female passenger slept in the coach; the men found places on the ground. About midnight, Isaac heard a rattlesnake. It terrified him, "in fact so much that I ran and jumped on top of the stage coach and scared the woman into hysterics." She thought the wagon was being attacked by Indians. Isaac stayed on the roof the rest of the night.

The Abilene-bound coach arrived the next day, as expected, and soon Isaac found himself skimming over a sea of wildflowers. Cartographers of the day called this the Great American Desert, but to Isaac it seemed they had gotten it wrong, for here was "a carpet of flowers such as words will not describe. The flowers rolled in the wind like vari-colored waves." Flowers north, south, east, and west—"the most beauti-ful vision in nature my eyes have ever beheld."

This did not last.

THE SKY TURNED cloudless and blue, the prairie brown. The flowers died. The Concho River went dry, although underground flows some-how kept portions of the bed flush with water and fish. The weather showed itself prone to fits of violence. A tornado followed him along a road. A "blue norther" caught him in the midst of a hunting trip and dropped temperatures from hot to freezing in minutes. He experienced heat like nothing he had known before. During a visit by the freak dragon winds that periodically blistered the Texas plains he recorded a temperature of 140 degrees Fahrenheit.

One evening in mid-August he was walking toward town along his usual route, crossing the footbridge over the riverbed, when he heard a roar from somewhere far upstream. Not thunder. The roar was continu-ous, and got louder. He saw a carriage carrying a man and two women descend into the riverbed at a point where wagons and horsemen often crossed. An escarpment of water that Isaac estimated to be fifteen or twenty feet high appeared beyond the carriage. Isaac began to run. The water caught the carriage broadside and ripped it from the soil. Isaac reached the other side of the riverbed just as the water surged past him, the carriage tumbling like a tree stump in a spring flood. The wagon passed. Rescue was impossible.

His heart racing, Isaac looked upstream. Men had gathered and with their bare hands were plucking fish from the water. Large fish. As Isaac walked toward the men, he saw a fish two feet long drift slowly by. He moved closer. The fish did nothing. He reached for the fish. It kept still.

Isaac thrust his hands into the water, and two things happened: He caught the fish; he froze his hands.

It was August in Texas but water had abruptly filled the riverbed and this water was the temperature of a Tennessee creek in January, so cold it paralyzed fish.

But where had the water come from? Isaac scanned the skies for the rolling black-wool cloud typically raised by blue northers, but saw nothing.

Days later, townsmen recovered the bodies of the carriage driver and his two female passengers.

A week later, the mystery of the ice-water flood was solved.

Visitors from the town of Ben Ficklin fifty miles up the Concho came to San Angelo and reported that a monstrous hailstorm had struck about ten days earlier, the day of the flood. The storm discharged stones the size of ostrich eggs that killed hundreds of cattle and fell in such volume they filled erosion gulches and piled to depths of up to three feet on level ground. The ice melted quickly.

For Isaac, this was explanation enough. The deadly flood was the downstream flow of flash-melted hail. He wrote an article on the incident for the weather service's *Monthly Weather Review,* edited by Cleveland Abbe. To Isaac's "surprise and chagrin," Abbe rejected the article on grounds it was too far-fetched to be believed.

The rejection stung. Isaac had been there when the flood came through. He saw the fish. He had thrust his hands in the ice-cold water. The shock of it on that August day in Texas was embedded in his brain.

Isaac could not let it go. Hail became a transient obsession. He tracked down reports of monster hail from all over the country. It was true, he wrote, that no one previously had reported a hailstorm so big as to produce a river of fish-paralyzing ice water, but on June 30, 1877, hailstones as large as oranges killed ponies at Yellowstone Valley, and on June 2, 1881, in White Hall, Illinois, hailstones the size of goose eggs piled to twelve inches deep, and on June 12, 1881, hailstones as large as a man's fist fell on three counties in Iowa and piled to depths of two or three feet, and on June 16, 1882, hailstones up to seventeen inches around and weighing two pounds fell at Dubuque, Iowa.

Which was Isaac's loyal, obedient, oblique, three-cushion way of stating that the great Cleveland Abbe had been wrong to reject his paper. Isaac was nothing if not credible, and did not like having his credibility challenged.

ISAAC FELL IN love.

The Signal Corps had moved his station to Abilene where Isaac began attending the city's Baptist church, led by Pastor George W. Smith. He was struck by the beauty of the music, and more to the point, by the beauty of the young organist who produced it. The woman was Cora May Bellew, a niece of Pastor Smith's who was living in the pastor's house.

"She was a beautiful, brilliant and cultured girl," Isaac wrote. "She had more attraction for me than any woman I had ever known."

He wooed her, won her, and, on March 17, 1887, married her. He remained true to his belief that one's time should be used efficiently, an ethos that Frederick Winslow Taylor soon would bring to American industry. An inefficient man, Taylor said, was like "a bird that can sing but won't sing."

Isaac could sing, and did. On December 10, 1887, after just eight and a half months of marriage, Cora May gave birth to a daughter. The Clines named her Allie May.

THE CLAMOR TO reform the weather service continued to grow. Demand for better and more useful forecasts intensified. Until the creation of the weather service, individuals had relied on their own meteorological savvy—and assorted almanacs, crackpots, and backwoods lore—to produce their own forecasts of the weather, just as they produced their own soap, bread, and clothing. But America as a whole was shifting rapidly toward a consumer culture in which remote factories produced the things families needed. Now a farmer could get a daily report from the Weather Bureau. "In the past the man has been first," Frederick Taylor wrote, "in the future the system must be first."

But was the system up to the task?

The weather service needed a hero, and got one. On January 16, 1887, Gen. Adolphus W. Greely took over as chief of the Signal Corps. He was by now one of the most famous men in America, albeit famous for having barely survived the failure of his 1881 expedition to Lady Franklin Bay in the Arctic, which left him marooned until his rescue in July 1884 by Capt. Winfield Scott Schley of the U.S. Navy, whose daring expedition made him a celebrity as well.

Captain Howgate, the embezzler, was still at large. Congress launched a formal investigation of the weather service. To gauge just how far the service had fallen, General Greely dispatched inspectors to weather stations around the country. In Greely's first year, he dismissed one hundred employees for all manner of offenses, including some that suggest that weathermen of the day were not drab bureaucrats who spent their lives watching mercury rise and fall. He fired one New England observer for indulging his passion for photography on bureau time. The observer turned the office into a studio where he photographed nude young women.

A fondness for extended fishing trips caused the head of the Rocky Mountain district to engage in some long-range forecasting. He would create a week's worth of weather observations, then unload them at the telegraph office with instructions to the operator to send them one by one over the following week. This worked fine, apparently—a testament either to the consistent character of Rocky Mountain weather or the observer's real forecasting savvy—until one of Greely's inspectors dropped in without warning. Finding the office vacant, the inspector went to the telegraph office and there discovered a neat stack of timed and dated weather reports awaiting transmission.

An observer in the Midwest turned out to be a compulsive poker player. Desperate for cash, he hocked the station's instruments. He took his daily readings at the pawnshop.

On January 21, 1888, while Isaac was still at Fort Concho, one of Greely's inspectors walked into the Galveston station. At the time, it occupied the third floor of a building that served as the city's police station and courthouse. The inspector, Lt. J. H. Weber, arrived at 1:00 P.M., and was greeted by Private E. D. Chase, the soldier then in charge. Lieutenant Weber checked the barometers with a plumb line to see if they were standing vertically. He checked whether they had enough mercury and if air had infiltrated their vacuum tubes. He reviewed the station's wind-signal record book and its expense book, and evaluated the performance and appearance of each man assigned to the station.

He did not like what he saw. He had not liked much of anything since the moment he arrived. Above all he did not like Private Chase.

The barometers were filthy. Lieutenant Weber had to clean them just to read the scales. Galveston merchants and agents of the Cotton Exchange complained loudly of neglect. Noted Weber, "They hardly look at the local office for information but depend mostly upon St. Louis and New Orleans papers for weather news." The station itself, he wrote, was in "execrable" condition. "Gentlemen should not be compelled to occupy quarters in which one would not kennel a well-bred canine."

The blame for this he laid entirely at the boots of Private Chase. "This man should be discharged for his miserable work while in charge here," Lieutenant Weber wrote. "He is not fit to remain in the service."

AND THEN CAME Monday, March 12, 1888: The Signal Corps's forecast for New York City predicted "colder, fresh to brisk westerly winds, fair weather."

What New York got was the Blizzard of '88. Twenty-one inches of snow fell on the city. Two hundred New Yorkers died. Nearly four feet covered Albany. The storm killed four hundred people throughout the Northeast.

This did not help. Not at all.

Isaac Cline was twenty-seven years old. He had a kind smile and welcoming manner, but a backbone like a frigate's mast and a capacity for heroic amounts of work. He was exactly the kind of man the Signal Corps saw as its salvation. In March 1889, General Greely ordered him to take over the failing Galveston station and, further, to establish the first Texas-wide weather service.

Isaac stepped from his train into a neat, well-ordered place, with alphabet streets running east and west, numbered streets running north and south. He had grown accustomed to the stark greens and grays of the sagescape that surrounded Abilene. The sudden blue of Galveston cooled his mind. He was struck, as all visitors were, by how flat the city was, so close to sea level as to produce the illusion that ships in the Gulf were sailing on the streets.

Avenue B, he quickly learned, was more commonly called the Strand. The Wall Street of the West. It sliced across the northern edge of the city just below the arc of wood and iron that formed the wharf front. The downtown streets were paved with flush-hammered wooden blocks and walled by knee-high curbs. Drays, sulkies, landaus, and victorias, with calash tops raised against the sun, eased along behind cautious head-down horses picking their way among the uneven seams. Each hoof struck the pavement with the thud of a mallet against wood, evoking the earscape of a building under construction. The clatter reinforced the aura of enterprise and industry.

Where Abilene had been a rude new town still redolent of fresh-cut wood, Galveston had substance. The size of its buildings and the obvious care invested in their construction betrayed the city's ambition to become something much bigger. Even in its hedonic infrastructure, Galveston displayed grand aspirations. The city had five hundred saloons, more than New Orleans, a city not exactly known for banking its fires. Galveston's poshest whorehouse was situated right behind its richest men's club, the Artillery Club, which barred women except for an annual ball and the occasional coming-out party of a member's daughter. The city's most disreputable block was Fat Alley, between 28th and 29th. In Galveston alcohol was blood, but one could also gamble, acquire love, and lose oneself in an opium mist.

The city exhibited a rare harmony of spirit. Blacks, whites, Jews, and immigrants lived and worked side by side with an astonishing degree of mutual tolerance. Through the Negro Longshoremen's Association, Galveston's black population controlled wharf labor and enjoyed a standard of living higher than almost anywhere else in the country. The immigrant influence was obvious. At the heart of town, Isaac found the Garten Verein, or Garden Club, built with money pooled by the city's German residents, who accounted for one-third of the population. It was a large, octagonal dance pavilion with pilasters, balustrades, and a central cupola, set in a park that included a bowling green, tennis courts, even a small zoo. Women could not smoke or wear rouge or lipstick on its grounds. But they could dance. In this staunch, straight-backed time when a man could not weep and a woman could not smoke, there was always dancing.

Galveston was too pretty, too progressive, too prosperous—entirely too hopeful—to be true. Travelers arriving by ship saw the city as a silvery fairy kingdom that might just as suddenly disappear from sight, a very different portrait from that which would present itself in the last few weeks of September 1900, when inbound passengers smelled the pyres of burning corpses a hundred miles out to sea.

· · ·

IT WAS NOT enough for Isaac to do merely what General Greely asked of him. He saw in his transfer to Galveston "great opportunities for the utilization of my recreation time." Although his colleagues might have been inclined to ask, *what* recreation time?

On August 24, 1889, his second daughter arrived. He and Cora named her Rosemary. They hired help, most likely. Everyone did. But a baby was still a baby. There were diapers but no washing machines. The nights were hard, the days tiring. As for Isaac's work life—the Galveston office was in disarray. Isaac was under orders not just to fix it, but also to start the new Texas-wide weather service. For most men, all this would have been quite enough. But in 1893 Isaac joined the faculty of the University of Texas medical school, based in Galveston, as an instructor in medical climatology, and during the year delivered thirty lectures on topics ranging from the fundamentals of measuring barometric pressure to the role of climate in pneumonia, malaria, and yellow fever. He also enrolled in Add-Ran Male and Female College, today's Texas Christian University, and began studying toward a doctorate in philosophy and sociology. He taught the young men's Sunday-school class at the First Baptist Church.

He quickly turned the Galveston office into a showpiece. On November 13, 1893, an inspector named Henry C. Bate paid a visit to the Galveston office, the first inspection since the transfer of the weather service in 1891 to the Department of Agriculture, which formally named it the Weather Bureau. Isaac, Bate wrote, "was exceedingly popular with everyone . . . The service has few such men in the field—none better." Bate provided the underlining.

By then, Isaac's brother, Joseph, had joined the bureau. Unlike Isaac, he had drifted toward weather. He taught school in Mount Vernon, Tennessee, for twenty-five dollars a month but quit to move to Galveston to become a salesman, or "drummer," for a printing company and quickly earned a reputation as being just about the only salesman in town who did not drink. He earned sixty dollars a month, but Galveston was a lot more expensive than Mount Vernon and he soon found he was saving less money. He joined a locomotive machine shop operated by the Gulf Colorado Railroad, but remained for less than two months. The fact Isaac hired him was evidence that for the moment the men were still close, still friends. At the time of Bate's inspection, Joseph was twenty-two years old and earning $840 a year, his best salary yet. Bate gave him a total score of 8.8, but noted his penmanship was "somewhat difficult."

In his concluding remarks, Bate wrote that the Galveston force was overtaxed and badly served by headquarters. "I don't think there is a station in the United States that gives out near the amount of information daily and weekly as this, and I am quite sure there is none where the value of the Service and this information is more genuinely appreciated than here." Yet few stations, Bate wrote, "are so poorly provided with office comforts and facilities—I hope the Chief will give this matter his favorable consideration."

A NEW CHIEF took over the bureau, Mark W. Harrington, the former editor of a meteorological journal. He continued Greely's campaign to reduce public skepticism about the bureau's ability to do much beyond simply recording changes in the weather. At the time of Harrington's appointment, Isaac wrote, "weather forecasting was nothing more than a listing of probabilities." Even something as basic as predicting the temperature twenty-four hours in advance was considered so likely to result in failure and public ridicule that the bureau forbade it. This prohibition frustrated Isaac Cline. He believed he understood the weather. He understood the rippling of isobars across the plains. Weather could be strange, but never so strange as to elude scientific explanation. Isaac had experienced tornadoes, hailstorms, freakish floods, and dragon winds. He understood them the way a parent comes to understand a difficult child.

Chief Harrington gave him a chance to prove it. In September 1893,

Harrington launched a competition, open to all, to find the best fore-casters in the Weather Bureau. The grand prize was a coveted profes-sor's position in Washington. The first step required contestants to write a paper, three thousand words or less, on the topic "Weather Fore-casts and How to Improve Them" and to submit this by December 1, 1893. Each contestant was to mail his paper under a false name to avoid prejudicing the three-man panel of judges, but seal his real name inside an attached envelope. Harrington received thirty entries. One came from Isaac. But Joseph, still an apprentice weatherman and nine years Isaac's junior, also submitted an entry. The rivalry intensified.

THE THREE JUDGES in Harrington's contest selected the ten best papers and invited their authors to Washington for the next phase of the com-petition, in which the finalists would take a written examination and spend two weeks testing their forecasting skills against those of their fel-low contestants.

Harrington sent two letters to Galveston. The first arrived Christmas Day and informed Isaac that he had placed among the top ten; the sec-ond told Joseph he had failed to make the cut.

The Galveston *News* applauded Isaac. "While those interested in the weather service work in Texas wish him success they would regret to see him called to other fields of duty, as his place here would be a hard one to fill." The *News* made no mention of Joseph.

Early in January 1894, Isaac went to Washington. He placed fifth in the final competition, but insisted his grade was only "three-tenths of one per cent behind the winners." Two other contestants tied for first, one a balding, mustached man named Willis L. Moore, with whom Isaac developed a warm personal friendship. Moore and his opponent entered a runoff competition. Moore won, and received the Washington professorship.

Joseph clearly felt hurt by his failure to place among the top ten finalists. He believed himself to be the best forecaster in the Weather Bureau and for proof cited the fact his name was first on all but one of the lists put out every six months by the bureau's forecast verification unit, which checked each prediction for accuracy. In a later memoir, he never mentioned that Isaac also had taken the test. In fact, in all 251 pages Joseph barely mentioned Isaac at all, and then only in the most cursory way.

It was geneology, by then. Not love.

THE YEAR 1894 brought Isaac a third daughter, Esther Bellew, his baby. There was a bit of good news, too, for the Weather Bureau. Police at last caught up with Captain Howgate, the fugitive embezzler. It was about the only good news, however. Conflict continued to embroil the bureau. It faced a nation of skeptics, one of the most ardent being Secretary of Agriculture J. Sterling Morton, Harrington's boss.

Morton wanted to save money and did not think he was getting full value from the bureau's scientists, whom he believed to be far too well paid for the little skill they demonstrated in forecasting the weather. The previous year he had launched an attack on Cleveland Abbe. In singling out the bureau's brightest light, it was clear Morton was attacking the bureau as a whole.

Morton's assault began on June 16, 1893, when he wrote to Abbe asking him to prove his worth. "It seems to me that the disbursements of the Weather Bureau for scientists are altogether too extravagant."

To Abbe, this was a jolt. In a reply drafted the next day, Abbe wrote, "Nearly every real advance in the progress of the Weather Bureau since I entered it, January 3, 1871, has gone through the three following steps, viz., first I have suggested and urged it; next I started the work and showed how it ought to be done; finally I found the best man, or organized a system, by which the work should be carried on as a permanent feature."

Morton, unmoved, demanded that Abbe send him proof of all these accomplishments. Abbe sent him a thick package of reports.

Five days later, Morton notified Chief Harrington that he had decided to slash Abbe's annual salary by 25 percent, to $3,000 from $4,000, "with the understanding that proficiency in forecasting will be necessary for the continuance of his services and the perpetuation of his pay." The man in charge of gauging his proficiency was to be Major Dunwoody, head of the forecast-verification unit, and one of that all-too-common category of men who feast on boot polish and see the failures of others as stepping-stones toward their own success. Dunwoody had been one of General Hazen's most ardent critics, objecting at every opportunity to Hazen's investment in scientific research. He would turn up again years later, in Cuba, doing his best to obstruct the efforts of Cuban meteorologists to transmit warnings about the hurricane of 1900 as it advanced through the Caribbean.

Dunwoody was a snake, and Chief Harrington knew it. At last, Harrington lost his patience. In a letter to Morton dated April 30, 1895, Harrington wrote: "Dunwoody is a selfish intriguer and a source of discord in the Weather Bureau. I request that the President recall him."

Instead, Morton fired Harrington. On July 1, 1895, Morton replaced him with Isaac's friend and fellow contestant, Willis L. Moore, only thirty-nine years old but already a veteran of nearly two decades of service within the Signal Corps and the Weather Bureau. It was an appointment that would shape in dangerous ways the bureau's ability to respond to the 1900 storm.

Moore tightened headquarters' control over the bureau's far-flung empire. He insisted on even stricter verification of forecasts. Dunwoody's verification unit kept busy, and filed a report on each man to Moore every six months. To further sharpen the bureau's skill, Moore insisted every observer do practice forecasts for a location outside his own territory so that on any given day a number of forecasters would try predicting the weather for the same city. This generated a lot of tension, but Moore believed tension was good. The system, he told Congress, helped explain why Weather Bureau employees had to be committed to insane asylums more often than employees of any other federal agency.

He said this with pride.

Moore also made himself guardian of the bureau's moral health, and in this role claimed broad jurisdiction. Early in 1900, in the midst of rising anticigarette sentiment that condemned smoking not for killing people but for making them stupid, Moore banished cigarettes from the bureau's weather stations. The Christian Endeavor Union of Washington promptly congratulated him. Moore, greedy for any scrap of praise, replied that he personally had dismissed bureau officials "purely on the ground that their moral character was such as to bring discredit upon the Weather service." Smoking was a moral blight. "In several cases," Moore crowed, "we have been compelled to take action for the reduction or removal of observers in charge of station for indolence, forgetfulness, and failure to render reports promptly, where I was satisfied that shattered physical condition and mental impairment were due to the excessive use of cigarettes. The order will be obeyed."

Moore never missed a chance to burnish the reputation of the Weather Bureau or to boost his own political stature. War provided a prime opportunity. By early 1898, the nation's bloodlust was high. The explosion of the battleship *Maine,* its true cause a mystery, had sent the nation tumbling irrevocably toward war with Spain. Clearly America's most important weapon would be its Navy. "I knew," Moore wrote, "that many armadas in olden days had been defeated, not by the enemy, but by the weather and that probably as many ships had been sent to the bottom of the sea by storms as had been destroyed by the fire of enemy fleets."

He reported his concerns to James Wilson, who by then had replaced Morton as secretary of agriculture. Wilson arranged a meeting between himself, Moore, and President William McKinley. Moore spread out a map of the Caribbean featuring the tracks of past hurricanes. McKinley studied the map, then turned to the secretary. "Wilson," he said, "I am more afraid of a West Indian hurricane than I am of the entire Spanish Navy."

Moore proposed the creation of a hurricane-warning service with stations in Mexico, Barbados, and elsewhere in the Caribbean. McKinley approved. He told Moore, "Get this service inaugurated at the earliest possible moment."

For that instant, at least, Captain Howgate was forgotten, the Blizzard of '88 forgiven.

The establishment of these hurricane-listening posts was too weighty a task for rank-and-file bureaucrats. Moore chose only trusted officers of the bureau. For the West Indies network, he picked Dunwoody. For the Mexican stations, he chose Isaac.

It was during this Mexican venture that Isaac encountered his first hurricane—at sea, no less. For many people, it would have been the defining event of a lifetime, the story told and retold every Thanksgiving until the waves were taller than Pikes Peak, the winds strong enough to knock a man clear to Halifax.

For Isaac, however, it had a different effect.

THE WEATHER WAS hot and still, the Gulf smooth as mica, but now and then despite the lack of wind a great hill of water slid silently under the ship and levered it high above mean sea level.

The sky at the horizon turned copper. Isaac had never seen such color in the atmosphere. Could this, he wondered, be the "brick-dust" sky he had read about in mariners' accounts of tropical cyclones?

His fellow passengers were unconcerned. At breakfast, one hundred men, women, and children crowded the ship's dining room, "all in a jolly mood."

Soon the sky darkened. Rain hammered the deck. The wind, by Isaac's reckoning, accelerated to hurricane force. The ship rocked and pitched in heavy seas. At lunchtime, Isaac found himself alone in the ship's dining room. Seasickness and fear had felled everyone else. He prided himself on his resilience. He made a show of it, no doubt, just as he had at Fort Myer, where he had raced his horse as fast as he possibly could while the city boys hugged their mounts and cursed his soul.

The storm continued through the day. At dinnertime not even Isaac appeared in the dining room. "I was so sick," he wrote, "that I did not care if the ship went to the bottom of the Bay of Campeche."

The ship survived. Isaac survived. He had met the most feared of all meteorological phenomena, yet had lived through it with only a case of seasickness. The experience had to have influenced his appraisal of the survivability of hurricanes. On some level, perhaps, he came to believe that hurricanes were not quite as awful as Piddington, Redfield, and Dampier had depicted. Or he assumed that technology—in this case, the modern steamship—had stripped hurricanes of their power to surprise and destroy. Indeed, in that same hurricane season of 1898 a naval architect from Pleasantville, New Jersey, named Simon Lake survived a particularly intense cyclone off Florida by submerging his submarine to a depth below the influence of the waves, exactly as Captain Nemo had done thirty years earlier in *Twenty Thousand Leagues Under the Sea.* "Jules Verne," Lake wrote, "was the director-general of my life."

Against the hubris of the age, what was a mere hurricane?

As THE YEARS passed, Galveston got bigger and more glamorous. Its future as a deep-water port seemed assured. In May 1900, the Galveston

News published a plan for the "Improvement of Galveston," devised by Col. H. M. Robert, divisional engineer, U.S. Army. Robert, famous by now for his *Rules of Order,* proposed an elaborate plan that would fill in the wetlands surrounding Pelican Island in Galveston Bay to produce an expanse of land eight feet above sea level called Pelican Territory. A harbor channel was then to be dredged between the territory and Galveston Island, and this was to serve as a portal to a new harbor basin with a surface area of seven thousand acres. The plan promised sure victory over Houston in the race to dominate the Gulf.

It did not include a seawall.

STRANGE WEATHER CAME and went. One episode revealed an unusual characteristic of Galveston Bay, but its true significance was lost among the more obvious phenomena of the moment.

The winter of 1898–99 proved a savage one. On November 26, just ten years after the awful Blizzard of '88, a powerful gale, known ever since as the *Portland* Gale, blew off the Atlantic and brought another surprise blizzard to New York. It destroyed 150 vessels off New England and caused the deaths of 450 men, women, and children, including all 200 passengers of the 291-foot paddle steamer *Portland,* whose captain had believed he could outrun the storm. Two months later, a blizzard swept much of the South. Icebergs ten feet high flowed down the Mississippi past New Orleans. The sudden cold killed participants in the Mardi Gras parade. The blizzard even struck Galveston and piled snow on its beaches. Snowmen populated the Garten Verein.

At the Levy Building, the temperature sank to 8 degrees, by Isaac's measure.

Seven-point-five, by Joseph's.

The wind blew from the north at up to eighty miles an hour, with so much power it literally drove water out of Galveston Bay into the Gulf, to the point where portions of the bay bottom lay exposed. Joseph, out hunting geese, claimed he was able to *wade* a channel ordinarily traversed by ocean-going ships.

No one, however, seemed to grasp the implications of this: that so vast a body of water could be blown from its basin. There were many distractions, however. There was snow on the beach. Icicles jutted from the underside of the Pagoda. Galveston residents filled rowboats with benumbed fish. Thousands of other fish accumulated along the bay shore in a blue-silver fringe four feet wide and half a foot thick.

The fish died. As the air warmed, the scent of death became overpowering.

THE STORM

Thursday, August 30, 1900:

17 N, 59.3 W

ON THURSDAY, AUGUST 30, 1900, the storm was just off the eastern coast of Antigua, where Francis Watts, an agricultural chemist with the government laboratory in St. Johns, observed a falling barometer and curiously shifty winds. At 9:00 A.M., the lab's barometer recorded pressure of 29.96 inches, still in the normal range. By midafternoon, the pressure had fallen to 29.84.

"About 10 P.M.," Watts reported, "a thunderstorm sprung up to the S.W. and came up over the land, appearing to be most severe over the region S.W. of St. Johns Harbor and generally within a radius of 3 miles of St. Johns. It died away after midnight. While it lasted it was very severe; the lightning was brilliant and almost continuous, while the flashes were very quickly followed by loud peals of thunder."

Shortly before the storm's arrival, strange weather had settled on the island. The day was intensely hot, the sky rimmed with a reddish-yellow light. There was, according to the Antigua *Standard,* an "ominous" stillness.

GALVESTON

An Absurd Delusion

IN JANUARY 1900, a self-styled weather prophet, Prof. Andrew Jackson DeVoe of Chattanooga, Tennessee, issued a long-range forecast for the year in his *Ladies' Birthday Almanac*. He predicted that September would be hot and dry throughout the northern states. "On the 9th," he wrote, "a great cyclone will form over the Gulf of Mexico and move up the Atlantic coast, causing very heavy rains from Florida to Maine from 10th to 12th."

It was the kind of prophecy Isaac Cline loathed. He was a scientist. He believed he understood weather in ways others did not. He did not know there was such a thing as the jet stream, or that easterly waves marched from the coast of West Africa every summer, or that a massive flow within the Atlantic Ocean ferried heat around the globe. Nor had he heard of a phenomenon called El Niño. But for his time, he knew everything. Or thought he did.

On July 15, 1891, the Galveston *News* published an article Isaac wrote on hurricanes. It is a troublesome document, for it abrades the body of convenient truth that has accumulated over the last century regarding Isaac's role in preparing Galveston for the hurricane of 1900. It tells worlds about what Isaac must have been thinking that Saturday morning and about how accurately he appraised the signs of approaching danger.

Isaac was only twenty-nine, but the article read as if it were written by a much older man. Clearly Isaac already considered himself a weather sage. He wrote the article in response to a tropical storm that ten days earlier had come ashore near Matagorda about 120 miles southwest of Galveston along the downward arc of the Texas Gulf Coast. Hubris infused the text just as it infused the age. He wrote with absolute certainty about a phenomenon no one really understood. He called the storm "an excellent type" of cyclone.

He explained first how the earth's rotation, the equatorial trades, and the midlatitude westerlies combined to give the storm a parabolic track that began near the equator, arced toward the northwest, then curved back toward the northeast. This last turn "nearly always" occurred between the 75th and 85th meridians of longitude, he wrote. (The 85th meridian passes through Havana, the 75th through the Bahamas.) Thus, he argued, hurricanes could not as a rule strike Texas. To buttress this observation he noted that during the two preceding decades, some twenty West Indies hurricanes had crossed the southern coast of the United States, but only two had actually reached Texas. "The coast of Texas is according to the general laws of the motion of the atmosphere exempt from West India hurricanes and the two which have reached it followed an abnormal path which can only be attributed to causes known in meteorology as accidental."

The article exudes an unmistakable scent of boosterism reminiscent of the immigrant come-ons published by the railroads. Clearly he understood how much was at stake in the race between Galveston and Houston, and that Galveston's promoters would not be pleased to read that the city lay in harm's way. He argued that if anything the coast was "much less susceptible" to hostile weather. "No greater damage may be expected here from meteorological disturbances than in any other portions of the country." In fact, he wrote, the "liability of loss" was much lower.

When storms did break the rules, he argued, they tended to be weak creatures. "The damage from the storm of July 5, 1891, aggregated less than $2,000, and yet was of much greater intensity than the average of these storms; and in fact no damage worthy of notice has been experienced along the Texas coast from any of these storms except those of 1875 and 1886 and in each of these two cases the loss of property aggregated less than that which often results from a single tornado in the central states."

These two exceptions were hurricanes that struck the town of Indianola, a prosperous port 150 miles southwest of Galveston on Matagorda Bay. By Isaac's analysis, the two hurricanes were accidents. Atmospheric freaks. But Isaac failed to grasp, or deliberately ignored, the true significance of the hurricanes, and what they did to Indianola. He focused on property damage. "The single tornado which struck Louisville, Ky., March 27, 1890, destroyed property of greater value than the aggregate of all the property which has been destroyed by wind and water along the Texas coast during the past twenty years."

Isaac had to have recognized the misleading impression this argument would conjure in readers' minds, unless of course he simply did not know what really happened in Indianola during those two storms.

For nowhere does he mention lost lives.

THE FIRST STORM struck Indianola on September 16, 1875. Gale-force winds had come ashore the previous day and gained velocity throughout the night. By 5:00 P.M. on the sixteenth the wind was blowing at eighty-two miles an hour. The wind continued to strengthen until by midnight, according to Sgt. C. A. Smith, the Signal Corps observer on duty, "it must have been fully 100 miles an hour."

The storm raised an immense dome of water and shoved it through Indianola, pushing the waters of the Gulf and Matagorda Bay inland "until for 20 miles the back country of prairie was an open sea." Residents fled their homes in boats and gathered in the town's strongest buildings. Shortly after midnight, Smith reported, the tide changed. The survivors believed the worst was over. "This evidence of abatement was hailed with shouts of joy, and was confirmed in a few minutes by the action of the wind, which gradually backed to the north and northwest."

Their joy was premature. The wind again began shoveling water, this time back toward Matagorda Bay, and created an "ebb surge," a mesoscale version of what happens on any beach when water brought ashore by a wave rushes back out to sea, undermining anything in its way. "The tide now swept out toward the bay with terrific force, the wind having but slightly abated, and it was at this time that the greatest destruction to life and property occurred. The buildings remaining had been so loosened and racked by northeast wind and tide that the moment the tremendous force was changed in a cross-direction dozens of them toppled in ruins and were swept into the bay."

The initial storm surge had poured into Matagorda Bay over the course of eighteen hours. It exited in six.

The devastation was stunning. "Fully three-fourths of all the buildings had entirely disappeared from the scene, and of those remaining, a large part were in utter ruins," Smith wrote. "Many of those remaining had been swept from their original foundation—some but a few yards, others several blocks."

The storm killed 176 people. Compared with the death tolls of the great Bay of Bengal typhoons, this raw total did not seem like much. But Gen. Adolphus Greely, who visited Indianola six months after the storm, estimated the death toll amounted to one-fifth the city's population. The storm left a schooner high and dry five miles inland and killed fifteen thousand sheep and cattle. All this, Greely observed, despite the fact that Indianola occupied a sheltered niche on the Texas coast fourteen miles from the Gulf and behind a broken plume of barrier lands that might have been expected to blunt the force of any oncoming storm. Even six months afterward, the damage was obvious and vivid. The hurricane had destroyed not only the superficial structures made by men, Greely found, but also God's own topography. "The striking physical changes were the formation of a large lake in the rear of the town and the plowing of numerous bayous inland, five connecting across the solid land of an elevation ranging between 10 and 20 feet above the level of Matagorda Bay, on which the town was built. One of these bayous was nearly 20 feet deep at the time of my visit."

Indianola was proud of its port and believed it could be restored to its former prosperity. Its residents chose to rebuild.

THE SECOND HURRICANE arrived on August 20, 1886. "The water in the bay commenced to rise rapidly," according to the Signal Corps account of the storm. The wind destroyed the service's weather station, where falling timbers killed the resident observer, I. A. Reed, as he tried to escape. "A lamp in the office set fire to the building and, although rain was falling heavily, it was burned, and also more than a block of buildings on both sides of the street."

The wind raised storm and ebb surges even more destructive than those of 1875. "The appearance of the town after the storm was one of universal wreck. Not a house remained uninjured, and most of those that were left standing were in unsafe condition. Many were washed away completely and scattered over the plains back of the town; others were lifted from their foundations and moved bodily over considerable distances."

The storm caused such thorough destruction, and killed so many residents, the survivors abandoned the town forever.

AT FIRST, GALVESTON'S leading men seemed to grasp the significance of the Indianola storms. Anyone who looked at a map could see that Galveston was even more vulnerable to destruction than Indianola. It had no picket of barrier islands to shelter it, no buffer of mainland prairie. The city faced the Gulf head-on.

Six weeks after the second Indianola storm, a group of thirty prominent Galveston residents calling themselves the Progressive Association met and resolved to build a seawall. This was the same group that led the fight for federal money to turn Galveston into a deep-water port. The city's engineer, E. M. Hartrick, went so far as to draft plans for the wall. He proposed "a dike ten feet high extending completely around the island, except for the north side. There, the wharves were to be raised to form the dike." The city's *Evening Tribune* endorsed the plan. "When men such as these say that work on seawall protection should be commenced at once and pushed to completion, the public can depend upon it that something tangible will be done—and that without unnecessary delay."

The state eventually did authorize a bond to pay for the work. "But," engineer Hartrick wrote, "this was some months after the flood, and by then the attitude was, Oh, we'll never get another one—and they didn't build."

If Galveston had any lingering anxiety about its failure to erect a seawall, Isaac's 1891 article would have eased them. It was here that he belittled hurricane fears as the artifacts of "an absurd delusion." He was especially confident about storm surges. Galveston would escape harm, he argued, because the incoming water would spread first over the vast lowlands *behind* Galveston, on the Texas mainland north of the bay where the land was even closer to sea level.

"It would be impossible," he wrote, "for any cyclone to create a storm wave which could materially injure the city."

PART II

The Serpent's Coil

THE STORM

Spiderwebs and Ice

THE STORM ENTERED the Caribbean Sea early on Friday morning, August 31, in a confetti of sparks and thunder, with increased winds that raised from the sea patches of dense foam and streaks of spindrift. In the cloudlight of morning the sea was a dead gray scabbed with green. Rain began falling on St. Kitts, an island west by northwest of Antigua. What made this rain unusual was the fact it did not deplete the clouds overhead. The storm only got bigger.

As vapor rose through the clouds and began to condense, it deposited its moisture on tiny bits of airborne debris, ranging from submicroscopic "Aitken" nuclei to pollen, spiderwebs, volcanic ash, steamship exhaust, Saharan dust, even the pulverized ferrous salts of meteors disintegrated in the atmosphere. Somewhere over St. Kitts, a giant plume of water, ice, and aerosol debris rocketed through the troposphere getting colder and colder until it penetrated the stratosphere, where it entered a realm of new warmth caused by direct radiation from the sun. Suddenly the plume was colder than the air around it. It lost buoyancy. It arced against the hard blue of the stratosphere and fell back toward the earth.

This descending air met air still rising from below. Falling droplets met ascending droplets. The collisions formed bigger drops and the bigger they grew, the faster they fell. Now they overtook other falling droplets and grew bigger still. A raindrop four-hundredths of an inch in diameter falls at nine miles an hour; a droplet six times as large falls at twenty. Billions of droplets now got bigger and bigger until they achieved terminal velocities capable of propelling them all the way to the ground.

Under ordinary circumstances, the process of rain production depletes clouds. The "sink rate," or the rate at which water leaves a cloud, exceeds the supply of moisture arriving from the air and sea below, causing clouds to dissipate like ghosts returning to the afterworld. But hurricanes defeat this cycle. They use wind to harvest moisture and deliver it to their centers. As the wind races along the surface of the sea, it increases the rate of evaporation and captures spindrift and foam. The faster the wind blows, the more vapor it picks up and the more energy it transfers to the storm. The resulting surge of condensation and heat in the storm's core causes even greater volumes of air to rush into the sky. Pressure falls again. Wind velocities increase. The cycle repeats itself.

The result can be rainfall more akin to the flow from a faucet than from a cloud.

In 1979 a tropical storm named Claudette blew off the Gulf of Mexico near Galveston and deluged the town of Alvin, Texas, with forty-two inches of rain in twenty-four hours, still the U.S. record for sheer intensity. A Philippine typhoon holds the world's record, dropping 73.62 inches in twenty-four hours. Total accumulations have been higher, however. Ninety-six and a half inches of rain once fell on Silver Hill, Jamaica, over four days. That's eight feet. In 1899 a hurricane dropped an estimated 2.6 billion tons of water on Puerto Rico. Hurricane Camille, which came ashore on the Gulf

Coast in August 1969, was still flush with water two days later when it reached Virginia. With no advance warning from the Weather Bureau, it jettisoned thirty inches of rain in six hours. Hillsides turned to mud, then to an earthen slurry that flowed at highway speeds. In Virginia alone, 109 people lost their lives.

Camille's rain fell with such ferocity it was said to have filled the overhead nostrils of birds and drowned them from the trees.

Louisa Rollfing

Saturday, September 1, was a big day in the home of August and Louisa Rollfing, a day for serious celebration. August, the housepainter secretly identified as a deadbeat in the Giles directory, had managed at last to make the final payment on the family's treasured piano. The moment had a resonance beyond the purchase itself. The piano was, literally, an anchor. It was heavy and big; just moving it into the house had required a huge effort—it had to be lifted in through a window.

This was their seventh house in Galveston. The house was a rental like all the others, but the piano made it feel more permanent, and Louisa badly needed such a sign. She was tired of moving. At each new address she had thrown herself into the task of making old and worn rooms look not only new, but as if they belonged to people of wealth.

She and August had been through so much turmoil, both individually and together. Both had come to America from Germany, August first at the age of one and with tragic bad timing. He and his parents arrived just as the Civil War began. His father, William, was promptly drafted, and just as promptly killed.

Louisa came to America much later, as a young woman. She had lived on an island in the North Sea, but had grown restless. There was so much talk of America. It started when a man named Daniel Goos came back to the island to visit family and told everyone of his big sawmill in a place called Lake Charles, Louisiana, where he had a wife and children and a large home. He needed more workers. He offered to take sixty people back with him to America. He would guarantee them jobs and advance them money for their passage. Many people Louisa knew went with him and they in turn sent for brothers, sisters, cousins, and sweethearts, until it seemed as if everyone was headed for America. A cousin and his wife now lived in Lake Charles and wrote often, each letter arriving at Louisa's house in a huge yellow envelope that her father placed in the window, a beacon of adventure that drew Louisa and her siblings at a run. "Just to hear the word America caused an excited feeling," Louisa wrote.

Each year more people left. The island got smaller and smaller. Her work as a housekeeper and companion for an elderly woman, Madam Michelson, made it positively tiny. There were days, it seemed, when Madam was the only person she saw. Louisa was lonely and dissatisfied and the idea of America crept deeper and deeper into her heart, until one day she simply resolved to go.

Her cousin sent her a ticket. She packed her things. Her confidence held until the night before her journey when she found herself lying awake, her heart racing, sleep an impossibility. "All at once I realized what it meant to leave everyone that was dear to me." The only thing that kept her going was the fact that her cousin would be waiting for her at the other side of the world. "I will never forget, when I saw Mother at the window, her big blue eyes filled with tears, smiling bravely—I had to run into the house and put my arms around her and kiss her again."

Louisa sailed on the North German Lloyd liner *Nurnberg,* accompanied by two young widows she had met in the emigrant hotel where everyone stayed before the voyage. Louisa was booked to travel in something called steerage, but had no idea what that meant. No one had told her she was supposed to bring her own blanket. Aboard ship, she and her new friends entered a great chamber "with nothing but wooden boxes on short wooden legs, with a thin mattress of straw on it, nothing else—they called them *beds!* Rows and rows of them. At the entrance was a great barrel, and we wondered for what?"

Louisa estimated that two hundred people occupied the hold, including complete families. "Oh I thought I would die. And cried bitterly. The two young widows felt just as bad as I did, and we shook hands that we would not be separated."

Soon after the voyage began, everyone got seasick, and the purpose of the great barrel became all too evident.

Louisa rebelled. She and her friends accosted an officer and demanded a more private place. They were women, after all. And single.

The officer had never before heard such a request, but agreed to look into the matter and later that day offered them a room in the stern, even to build them a partition for privacy—provided they could gather enough other single women to make the effort worthwhile. Louisa and her friends corralled thirty-four.

The journey to New Orleans and from there to Lake Charles took forty-two days. Her adventures began as soon as she arrived. She tried her first banana, and fell in love with it. She met her first black man. She was walking through a lovely stand of pine trees, when he appeared suddenly on the path ahead. "I got so scared that I just sat down, but he only said 'Good Day' and passed. *He did not kill me.*"

She caught the measles. "I got very sick," she said. "For a long time someone had to be up all night with me, and I did not even know it." She finally got out of bed six weeks later. When she looked at herself in the mirror, she saw that someone had cut off all her hair. She weighed only eighty-nine pounds, one-third less than when she had stepped off the boat. She had been beautiful. Now she was ugly. She was weak and vulnerable to other illness. A doctor advised that she move to a place with a healthier climate, perhaps Galveston.

Her train was halfway across one of the trestles that spanned Galveston Bay when she awoke and saw only water on both sides of her coach. She was terrified. She had not known Galveston was on an island and wondered how exactly she had ended up aboard a boat.

She was glad, later, that she had been unable to see the flimsy trestle. "I would have been scared even more."

In Galveston, Louisa took work as a housekeeper for a family named Voelker. On a Sunday visit to the home of Mrs. August Rollfing, the widow of a sea captain who had drowned in a storm off Galveston, she met Mrs. Rollfing's nephew, also named August. He was, Louisa confessed, "the nicest looking young man I had ever seen."

Not just handsome—but talented. He was a painter, and he played guitar and piano, and sang so beautifully. "He had a lovely tenor voice and I enjoyed it more than anything else in the world."

Some while later he proposed to her, if a mite obliquely. "Don't you think, Louisa, we could always be happy together, and that we should get married?"

It was a lucky thing that neither put much stock in omens.

One evening in November 1885, a week before their wedding, Louisa sat working on her wedding dress, a wonderful thing of gray cashmere with lace trim. She stopped work around midnight, folded the dress carefully, and brought it up to the room. "I wasn't even asleep when the fire whistle blew, and we saw a fire over at the north."

A powerful north wind was blowing—a blue norther—which quickly fueled the fire and blew sparks and large flaming cinders onto downwind homes. Another house caught fire. Then another. Louisa threw on some clothes, as did everyone else in the Voelker house, and all watched the blaze. No one thought the Voelkers' house might be in danger. The fires were still all so distant.

Butterflies of flame drifted through the sky. One moment the air was hot with radiated heat, the next, bitterly cold from the fierce north wind. A neighbor's house caught fire. Voelker climbed to his roof with a garden hose. Everyone else began hauling things out of the house. Louisa placed her trousseau in a trunk, which wound up on the sidewalk. The house caught fire. Trees caught fire. The trunk caught fire. Even Louisa's coat caught fire.

That night half of Galveston burned to the ground, and with it Louisa's trousseau—but, luckily, not her dress.

August and Louisa married on schedule. Even disaster could not dampen their spirits. "I can't imagine anybody happier than we were at that time," she recalled. "It took so very little to make us happy and contented."

During one of their many walks, they spotted a small white house for rent at 32nd and Broadway, and leased it the next day. It had a front porch, back porch, dining room, kitchen, bedroom, and a picket fence that surrounded the yard. Louisa threw herself into fixing it up. She bought a bed with a red canopy. She put up cream curtains with red tiebacks and made a red-and-white bedspread. She bought a large rug, lace curtains, and a hanging lamp with prisms and colored glass. She made a drape to cover a window in the living room that opened on the kitchen, and August decorated one side with a painting of flowers, fruit, and cupids. Soon the house was awash in rich colors touched with flowers and gold. "We felt as if we had heaven on earth."

Louisa sewed to raise extra cash, and took in more and more work. Was it too much work, she wondered forever after—too much for a woman pregnant with her first child?

Peter August was born April 8, 1888, months too soon. "He was just like a little doll and his little hand would lay in my hand." Her doctor told her to nurse the baby every two hours, but she could not. She was sick and weak and Peter August refused to nurse. "I did not have any experience, maybe he could have been saved if I had Mother near me."

Her son lived seventeen days.

For the funeral, Mrs. Voelker placed tiny white rosebuds over Peter's body and one rose in his hand. Louisa was too sick to accompany her baby to the cemetery. She watched as August placed the boy in a tiny white coffin and carried him to a carriage parked out front. It was days before Louisa could go to the cemetery. "When I came the first time they had his full name laid out with little white shells, and planted violets, and they were growing, for they came every day and watered them." August ordered the construction of a small wooden cross, then painted it himself. With ever so much care he painted his son's name and the dates of his life on the cross-spar, in gold letters.

"Now we had a place to go," Louisa remembered, many many years later. "It was our first sorrow."

She could not imagine loving her husband more.

Another child followed, "our little Helen," born fifteen months later. Louisa was fiercely proud and thought the child the most beautiful creature on earth, although in fact the baby was quite plump. "I can't see any *nose,*" August teased. Louisa was furious. At night she placed Helen in a baby buggy under a big, loosely draped mosquito net. "She looked like a little fat angel."

Two years later, another baby arrived, this one August Otto. The couple's landlady, a Mrs. Carville, came to visit the new baby, and saw for the first time all that Louisa and August had done with the house. She promptly raised the rent. August was irate.

They moved. Then moved again, had another baby—Atlanta Anna, or simply "Lanta"—and moved again. One thing after another forced them to leave houses into which Louisa had poured her soul. But the sixth house made it all seem worth it. "We found a very nice small two-story at such a reasonable price, that I could hardly believe it." Again she fixed it up, to the point where one Sunday afternoon an elderly couple, guests for dinner, arrived and wound up hopelessly confused, thinking they had come to the home of rich people, until Louisa, who had watched from the parlor window, gaily opened the front door and announced, "It *is* the right place!"

But it was not. Something was wrong. Someone else apparently had poured a soul into this house, but this soul had not yet departed. Evenings, after Louisa had put the children to bed, she sat alone until about ten o'clock, the time when August usually returned from rehearsing the amateur musicals in which he sang (sang, perhaps, alongside Isaac Cline, another tenor). She would settle in the living room to sew or read, but always she was seized by the same strange feeling. "It always felt as if something was looking over my shoulder; when I looked around there wasn't anything."

One night August felt it too. "I was alone," he said, "and still wasn't alone; there was something creepy around me."

They resolved to move yet again. "I was awfully disappointed," Louisa said. "Everything was so pretty and I was tired of cleaning and fixing."

They found another little two-story house, at 18th and O½, about ten blocks from Isaac's house and only two and a half blocks from the beach.

That summer—the summer of 1900—little August disappeared. It was a Sunday afternoon. The children were out playing. Louisa called them in for dinner. Helen and Lanta came, but not August. They had dinner, and still August did not come, and Louisa and her husband began to worry. The beach had always been an anxiety for Louisa, as it was for most parents in Galveston's beach neighorhoods. "I went east and August went west," she wrote. She walked the sand until exhausted, but did not find the boy. When she turned the corner onto her street, she saw that a crowd of children had gathered on the sidewalk in front of her house. She knew the worst had happened. She wanted to run to the house, but could not. Her limbs felt so heavy. She could hardly move.

She saw no one on the first floor. She climbed to the second and there found her husband. She said nothing, asked nothing. Soon her husband told the story—how little August and a friend had wandered to the beach, and walked and walked without realizing the distance. The walk back had taken forever.

Louisa served her son his dinner. Then the rest of the family, including young August, set out for an evening stroll. Louisa stayed back for a little while, in the warm light of dusk. She cried. And when she was done, she too walked to the beach and caught up with her family. It was a lovely evening, the sea so peaceful and edged in the gold of the setting sun, the mist blending all the blues and golds and the black and white of the hundreds of people who strolled also along the beach, none aware that for a few moments that afternoon she believed her heart broken for all time.

On Saturday, September 1, August made that last payment on the piano. Next, he resolved, he would find a piano teacher for Helen.

"If we had known what the future had in store," Louisa wrote, "we would not have had any pleasure in anything we did enjoy so."

Isaac's Map

AT THREE O'CLOCK in the morning, Tuesday, September 4, a lightning strike knocked out the incandescent-lamp dynamo at the Brush Electric Power plant in Galveston and cast the city's public buildings into darkness. The blackout showed how quickly people had grown dependent on electric lights, how willingly they abandoned the bad old days of gas jets, lamp oil, and kerosene.

At the police station, officers scrambled to find some means of lighting the station and its jail cells. A witness found an eerie scene: "A large assortment of miscellaneous lamps and lanterns shed faint gleams of light that were distressing to behold." The police had scavenged two calcium-carbide bicycle lights from the department's two patrol bicycles. Two old "bull's-eye" lamps dating to the 1870s "cast a flickering yellow ray of light within a radius of about eight inches." The police found and deployed three old railroad lanterns, which burned bright for ten minutes, then began to waver. The art of trimming and coaxing such lamps had been lost. The officers located the station's old gas jets, but these were in such poor shape they dared not light them. They did not possess any candles.

Isaac heard the first clap of thunder at 3:48 A.M., and later noted the time in the station's daily journal. He stayed up to listen, partly out of professional responsibility, partly because, like all meteorologists ever born, he loved thunderstorms. He walked onto his second-floor porch and there noted the occurrence of each electric burst, and how different the lightning was from that in Tennessee. In the knob country of his childhood it writhed across the sky in ruptured webs. Here it came in blue-white shafts, each spasm like the flare of flash powder from a photographer's trowel. In that instant, Galveston became an Arctic city of silver and black, a dying mariner's dream.

The loudest thunder occurred at 4:57 A.M., Isaac noted, the last at 5:20. The storm had come from the southeast, the direction of Cuba.

After breakfast, Isaac walked to the office. High above the warehouses along the wharf he saw thickets of masts and spars and the tall funnels of steamships. Some mornings, the varnish and brass caught the sun and made this tangle of line and wood gleam as if glazed by an ice storm. When the breezes were sluggish, smoke from coal-fueled steamships drifted over the streets in fat indigo plumes until the entire wharf seemed to smolder. One of the newest arrivals was the big British *Roma*, which had docked Sunday after a passage from New York. Its captain had the improbable name Storms.

Isaac's walk on Tuesday morning was especially pleasant, because the thunderstorm had dropped the temperature by a full seven degrees.

AT THE OFFICE, Isaac examined the 8:00 A.M. Washington weather map composed that morning by Theodore C. Bornkessell, the station's printer, using details telegraphed from headquarters. Bornkessell's graphic version included loopy isobars that linked areas of equal atmospheric pressure and dotted isotherms that did the same for temperature. Isaac sent a man to the Cotton Exchange to compose its large-scale version of the map. He might have sent his brother, Joseph, or Bornkessell, or a new man named John D. Blagden, on loan to ease the station's workload since the recent departure, in disgrace, of an assistant observer named Harrison McP. Baldwin. Baldwin, the Fort Myer clown, had come to work for Isaac a year earlier and quickly tarnished the station's reputation for accuracy. Throughout July and the first weeks of August 1900, error messages flowed from Washington to Galveston citing mistakes that Baldwin had made, and that Isaac was obligated to acknowledge and correct. The errors pained Isaac deeply. Chief Moore suspected Baldwin of far greater sins. He told Secretary of Agriculture Wilson he believed Baldwin had "fabricated" barometric readings—the highest of crimes. In mid-August, Moore put Baldwin on mandatory furlough, without pay. Baldwin left Galveston at 5:30 P.M., Monday, August 27.

Isaac was no doubt glad to be rid of Baldwin. The man had been a drag on performance and morale. It is likely, too, that Baldwin made fun of Isaac. Certainly Baldwin was given to pranks and poking fun. To a man like that, Isaac had to have been an irresistible target. With Baldwin gone, however, Isaac found himself short of hands. Moore promised him a junior observer named Ernst Giers, but at the last moment, for reasons Moore felt no obligation to explain, Moore rerouted Giers to Carson City, Nevada. The abrupt reassignment moved Isaac to a rare, if quiet, expression of complaint. He telegraphed Moore: "Giers not having arrived impossible to get along without experienced assistance in Baldwin's place."

Moore sent him John Blagden.

THE BUREAU'S MAPMAKER used colored chalk to compose the Exchange map. He noted pressures, temperatures, rainfall, and wind direction on a large blackboard painted with an outline of the country. That morning, Dr. Samuel O. Young, the secretary of the Cotton Exchange and an amateur meteorologist, came by to observe the process.

For the last week, Young had been keeping close track of the weather. Nothing in the official reports from the Weather Bureau's Central Office indicated that a tropical cyclone might be forming in the Caribbean, but Young believed the signs were there.

He stood quietly beside the mapmaker. There was something soothing about the *tap-tap-tap* of the chalk, as the mapmaker deftly noted wind speed in Chicago, temperature in New York, pressure over the Rockies. An *R* meant rain, *S* snow. An *M* stood for *missing*.

Under Moore, only disaster or downed telegraph lines made an *M* acceptable.

Tiny circles with arrows, like the symbols for male and female, soon covered the map. An open circle meant clear skies. A cross meant cloudy. The arrow showed the direction of the wind.

The mapmaker drew his isobars with assurance and grace, the chalk making a sound like skates on ice. He applied the dotted isotherms with special gusto, in Gatling bursts that turned his knuckles white.

Typewriters cackled. A telephone rang. Motes of chalk dust drifted in the gray light, like virga from a cloud.

Dr. Young paid special attention to the notations the mapmaker applied along the Gulf and Atlantic Coasts. "When the observations at Key West were recorded," Young wrote, "I saw that the barometer was low, that the wind was from the northeast and that the map as a whole showed pretty plainly cyclonic disturbances to the south or southeast of Key West."

There was no specific symbol on the map that indicated a tropical cyclone. Young deduced its presence from the unusual pattern of pressure and wind. He noted also the high-pressure zones that still lingered over the Midwest and Northeast. To him, the play of isobars and wind suggested a cyclone might be churning in the sea somewhere south of Florida, perhaps Cuba, and he said as much to the mapmaker.

"He agreed with me," Young wrote, "but said his office had received no notice of anything of the kind."

Suspicion

THERE WAS BAD weather in Cuba—*mal tiempo*. There was also bad blood. Willis Moore's passion for control had gouged a deep chasm between Cuban and U.S. meteorologists.

Moore and officials of the bureau's West Indies hurricane service had long been openly disdainful of the Cubans. It was an attitude, however, that seemed to mask a deeper fear that Cuba's own meteorologists might in fact be better at predicting hurricanes than the bureau. In August, Moore moved to hobble the competition once and for all. The War Department was then still in charge of Cuba, as it had been ever since the end of the Spanish-American War. Moore's chief liaison on the island was H. H. C. Dunwoody (now Colonel Dunwoody), the bureaucratic intriguer who had helped undermine Moore's predecessor, Mark Harrington. Through Dunwoody, Moore persuaded the War Department to ban from Cuba's government-owned telegraph lines all cables about the weather, no matter how innocent, except those from officials of the U.S. Weather Bureau—this at the peak of hurricane season.

It was an absurd action. Cuba's meteorologists had pioneered the art of hurricane prediction; its best weathermen were revered by the Cuban public. Over the centuries, storm after storm had come to Cuba utterly by surprise, until 1870 when Father Benito Vines took over as director of the Belen Observatory in Havana and dedicated his life to finding the meteorological signals that warned of a hurricane's approach. It was he who discovered that high veils of cirrus clouds—*rabos de gallo,* or "cock's tails"—often foretold the arrival of a hurricane. He set up a net-work of hundreds of observers, runners, and mounted messengers to watch for changes in the weather and spread the alarm. After Vines's death in July 1893, Father Lorenzo Gangoite took his place at Belen and likewise devoted his life to storm.

But the Weather Bureau under Willis Moore wanted hurricanes all to itself. After the war, Moore headquartered the Indies network in Havana. Dunwoody served as the bureau's senior representative on the island, but the man who actually ran the stations day by day was a bureau manager named William B. Stockman, the local forecast official for Havana, who saw the people of Cuba and the Indies as a naive, abo-riginal race in need of American stewardship.

"It was at first very difficult to interest the various peoples in the warning service," Stockman wrote to Moore in a voluminous June 1899 report on the Indies service's first full year of operation, "as the inhabi-tants of the islands are very very conservative and it is most difficult to get them to adopt any measures that radically differ from those pursued by their forebears, and forecasting the approach of storms, etc., and dis-playing warning signals or issuing advisory statements relative thereto, was a most radical change—the inhabitants being accustomed to hear of these phenomena only upon their near approach to a place or after it had passed in the vicinity."

It was as if Father Vines had never lived, and the Belen Observatory had ceased to exist. Eventually Belen's Father Gangoite discovered Stockman's remarks. By then, however, the corpses floating in the hot seas off Galveston had freighted Stockman's words with a brutal, unin-tended irony.

Stockman was a ponderous bureaucrat, given to writing immense reports about tiny things. When he filed his second annual report on July 31, 1900, even the professors and clerks at the Central Office rebelled, and these were men accustomed to levels of tedium that would have driven ordinary men to suicide. Internal memos flew from department to department, politely recommending that Stockman be muzzled. On August 15, Professor E. B. Garriott, one of the bureau's most senior scientists, wrote to the chief clerk: "I am loth to criticize the work of a man who has shown commendable zeal in the prosecution of that work. Nevertheless I am constrained to say that if the Official in Charge at Havana could curb a tendency toward verbosity and avoid iterations and reiterations in successive communications of matter that is irrelevant and immaterial to the subject heads, a great deal of time and labor would be saved both at Havana and the Central Office."

Willis Moore's recommendation was a bit less florid: "Kindly tell him to save himself much work."

In most other respects, however, Stockman was a good man to have in Havana. He shared Moore's obsession with control and reputation, as did the men Stockman placed in charge of the hurricane stations on outlying islands. Like Moore, Stockman worried about the damage likely to occur through the issuance of unwarranted storm alerts. In the Indies service, however, this concern took on a colonial cast. The poor, ignorant natives were too easily panicked. Restraint was the white weatherman's burden. It was paramount, he wrote, that the service avoid causing "unnecessary alarm among the natives."

He saw conspiracy everywhere. The Cubans, he believed, were trying to steal the bureau's weather observations to improve their own forecasts. He spent a good part of August 1900 investigating a man who called himself Dr. Enrique del Monte and claimed to be a professor at the University of Havana. In April, del Monte had published a well-received essay, "The Climatology of Havana," in the bureau's own *Monthly Weather Review*. Briefly, del Monte had even worked for Stockman. But now Stockman believed del Monte to be a fraud, perhaps even an agent of the Belen Observatory.

Stockman composed a nine-page letter to Willis Moore, dated August 10, which he devoted entirely to del Monte. He parsed del Monte's article. In the essay, the doctor had described his observatory and the shelter that housed its instruments, and told readers it was located on a particular train line in Havana. *Ah*—but no such observatory existed! Stockman checked. "The shelter described for the exposure of the thermometers exactly describes the structure used for said purpose by the Belen College Observatory."

Moore too suspected the Cubans, and believed a conduit for purloined weather intelligence ran between New Orleans and Belen. On August 24, 1900, W. T. Blythe, director of the bureau's New Orleans section, wrote Moore a letter that stoked his suspicions. He notified Moore that the College of Immaculate Conception in New Orleans received a copy of the national weather map every single day—the college simply dispatched a messenger to his office to pick one up. He did not feel he had the authority to refuse. He suspected, however, that the college then transmitted the contents by submarine cable to Belen. The real purpose, Blythe wrote, was "to enable the Belen College in Havana to compete with this Service."

It was all too much for Moore. Too clear. Moore instituted the ban on Cuban weather telegrams and halted all direct transmission of West Indies storm reports from the bureau's Havana office to its New Orleans station. The bureau even sought the help of Western Union. On August 28, Willis Moore, then serving as acting secretary of agriculture, wrote to Gen. Thomas T. Eckert, president of Western Union. "The United States Weather Bureau in Cuba has been greatly annoyed by independent observatories securing a few scattered reports and then attempting to make weather predictions and issue hurricane warnings to the detriment of commerce and the embarrassment of the Government service." He revealed his suspicions of the New Orleans connection. "I have reason to believe that they are copying, or contemplate doing so, data from our daily weather maps in New Orleans and cabling the same to Havana."

Moore closed the letter, stating, "I presume you have not the right to refuse to transmit such telegrams, but I would respectfully ask that they be not allowed any of the privileges accorded messages of this Bureau, and that they be not given precedence over other commercial messages."

To the Cubans, the cable ban was an outrage. "This conduct," wrote the *Tribuna* in Cienfuegos, "is inconceivable." Especially at the peak of hurricane season, "when everybody is waiting for the opinions and observations" of Cuba's hurricane experts. The newspaper cited in particular the reports of a meteorologist named Julio Jover. The cable ban, it cried, represented "an extraordinary contempt for the public."

The uproar took the bureau by surprise. Apparently Moore, Dunwoody, and Stockman expected the backward peoples of Cuba to accept the ban just as they accepted the daily rise of the sun. On Wednesday, September 5, as the storm of 1900 moved toward Havana, Dunwoody wrote to Stockman: "A very bitter opposition is being made both officially and through the newspapers, to the order prohibiting the transmission of weather bureau dispatches, by cranks on the island.

"I am not certain whether my position will be sustained by higher officials, but I have made the issue on the basis of good service. Of course, it will be necessary for you to furnish the press with good reliable warnings, in order to defend the stand I have taken."

Dunwoody stood firm, and for the moment prevailed. The War Department allowed the ban to continue.

Stockman and the observers in his network took special pains to avoid using the word *hurricane,* except when absolutely necessary or when stipulating that a particular storm was *not* a hurricane. They took what might be called a behavioralist approach to storms. They collected readings of temperature, pressure, and wind, and based solely on these, determined whether a storm existed or not. They sent clipped telegrams in a code that did not allow for conjecture or expressions of instinct, yet in their seeming precision produced the same sense of mastery over the weather that daily weather journals gave to men like Thomas Jefferson and George Washington. To Stockman, the tropical storm then making its way over Cuba was the sum exactly of its parts, no more and no less. And the parts did not add up to much. On Saturday, September 1, he released the bureau's evaluation of the storm to the *Diario de la Marina,* in Havana. "A storm of moderate intensity (not a hurricane) was central this morning east by south of Santo Domingo. . . . Fast steamers which sail today from Havana for New York will reach their destination ahead of the storm."

The Cubans took a more romantic view, a psychoanalytic approach, that was the product of the island's long and tragic experience. Nearly every Cuban alive had experienced at least one major hurricane. Cuban meteorologists had the same instruments as their American counterparts, and took the same measurements, but read into them vastly greater potential for evil. The Cubans wrote of hunches and beliefs, sunsets and forboding. Where the Americans saw numbers, the Cubans saw poetry. Dark poetry, perhaps—the works of Poe and Baudelaire— but poetry all the same.

They were wary from the start. On August 31, Julio Jover reported his assessment of the atmosphere to *La Lucha* in Havana. Barometric pressure had begun to rise, he noted—but he saw no comfort in the fact: "This, far from proving to us that the indications of a cyclone are vanished, reaffirms our opinion of the unstable equilibrium of the atmosphere, and therefore of the increase in energy of the center of low [pressure] which is over the Caribbean Sea."

The next day, Belen's Father Gangoite released to *La Lucha* his view that the storm, while at the moment a small one, appeared to be "a cyclonic disturbance in its incipiency. . . . This kind of storm sometimes produces heavy rain over this island, and acquires greater energy as it moves out over the Atlantic."

Father Gangoite was right about the rain—

Between noon and 8:00 P.M., Monday, September 3, Santiago received over 10 inches. The rain kept coming. By Friday, the total reached 24.34 inches, enough vertical flow to fill a claw-foot bathtub.

—but Gangoite was right, too, about the energy.

Captain Halsey's Choice

AT 9:20 A.M. Wednesday, Captain T. P. Halsey of the steamship *Louisiana,* then moored in New Orleans, ordered his crew to cast off the main hawsers and make for the Gulf. He saw a red-and-black storm flag rippling in the wind at Port Eads, Louisiana, but believed he had nothing to fear. Nothing in the reports from the Weather Bureau indicated conditions capable of threatening a modern steamship—there was no reference at all to gales or cyclones, no indication whatsoever that the storm could be a hurricane, or even had the potential to become one.

And if a cyclone did materialize, so what? He had survived eight so far.

The Weather Bureau's reluctance to use words like *hurricane* and *cyclone* inadvertently reinforced the bravado of sea captains like Halsey. Many mariners still believed that whether a ship encountered a storm or not was largely a matter of chance, so why worry? It was an ethos of resignation born of the frequency with which hurricanes took the ships and lives of even the best captains. Wrote Piddington, in a late edition of his *Sailor's Horn-Book,* "we must expect to find many 'of the old school' who do not like 'new-fangled notions;' many who 'do not like to be put out of their way;' many who 'think the old plan is good enough;' and that 'hit or miss, for luck's all,' is quite enough with a stout ship and a good crew." Modern technology helped perpetuate this ethos. Steel and steam produced ever-stouter ships. Engines reduced the worst storm hazards—the loss of control after sails were furled, the imbalance imparted by suspending tons of timber, canvas, brass, and rope high above a ship's deck. Technology was an elixir for last-minute qualms.

The *Louisiana* entered the main body of the Gulf at 5:22 P.M. Halsey's barometer read 29.87 inches. Winds were from the east-northeast, the top-left quadrant of a cyclone. The storm itself was moving toward the northwest. If Halsey had held one of Henry Piddington's transparent storm cards on a chart over his position, he would have seen that his ship now lay directly in the cyclone's path.

To be concerned, however, he first had to know that a cyclone even existed. All Halsey knew was that a nondescript tropical storm was at that moment arcing north into the U.S. mainland and soon would cross into the Atlantic.

To Halsey, it was fine, brisk day to be at sea.

A Matter of Divination

ON WEDNESDAY MORNING the storm rumbled into the Straits of Florida just north of Cuba and promptly confounded the Weather Bureau's forecasters. Willis Moore and his professors believed the storm would now move north. To them, the storm appeared to have begun a long turn or "recurve" that would take it first into Florida, then drive it northeast toward an eventual exit into the Atlantic. No real evidence supported this projection. It was merely what the latest iterations of the Law of Storms predicted and what the bureau's scientists expected based on the little that was known about tropical cyclones. In the age of certainty, at the gateway to the twentieth century, the expected was as good as fact. To turn was every storm's destiny.

Shortly after noon on Wednesday the Central Office telegraphed a report to New Orleans that the storm "probably will be felt as far north as Norfolk by Thursday night and is likely to extend over the middle Atlantic and South New England states by Friday." So far, the report said, "the storm has been attended only by heavy rains and winds of moderate force."

The report contained some excellent news—the storm would "terminate the period of high temperature which has prevailed east of the Mississippi."

IN HAVANA, WEDNESDAY, Julio Jover sent an 8:00 A.M. dispatch—by mail—to *La Lucha*: "We are today near the center of the low pressure area of the hurricane."

Again, that dreadful word.

When William Stockman read Jover's report, he surely laughed. He cut the report from the newspaper and affixed it to a special form designed by the Weather Bureau to help station chiefs collect praiseful articles from the nation's newspapers and forward them to Moore. Stockman saw Jover's report as further justification for the telegraph ban—it was another example of alarmist forecasting by the Cubans, who seemed to care more about drama and passion than science. Stockman did not consider the storm worthy of much further attention.

THE STORM AND its expanding cyclonic system now influenced a territory covering a million square miles of ocean and began to shape the weather in the southern United States. In Tampa, telegraph wires whistled. Winds reached twenty-eight miles per hour. In Key West, the barometer fell to 29.42 inches, the lowest level yet reported. The wind came from the northeast and accelerated to forty miles an hour, a true Beaufort gale.

Wednesday evening, however, the wind in Key West abruptly weakened. Its velocity dropped to six miles an hour, barely a breeze. Later that night, it began accelerating again, this time from the south.

The bureau's forecasters believed the sudden easing of the wind and the attendant change in direction meant the center of the storm had passed over or near Key West, and saw this as confirmation of their belief that the storm would soon be traveling up the Atlantic coastline. Once again, they tailored fact to suit their expectations. They knew just enough to believe they had nothing to fear.

But the storm did not go north.

The bureau had missed the true meaning of the wind shift at Key West. Here was an area of calm immediately adjacent to a zone of gale-force wind, in a storm that had just crossed the great mass of Cuba without losing any of its size or energy or its ability to produce biblical volumes of rain. No one knew it at the time, but the conditions at Key West provided the clearest evidence yet that the storm's architecture was changing.

At the storm's center, centrifugal force had come to play—the same force that flings children off the rims of playground carousels. The winds spiraling toward the storm's center now traveled at such a high rate of speed, they began to generate centrifugal force that sought to push them back out again. Where the inrushing and outpushing forces balanced, the winds began to form a circle, a gigantic carousel over the ocean.

This storm was about to open its eye.

THE NEXT MORNING, Thursday, at 6:00 A.M., William Stockman sent a dispatch that placed the storm 150 miles north, by *east,* of Key West.

It was a grave mistake, for it colored the expectations and perceptions of the bureau's Central Office at a critical point in the storm's journey. Stockman's guess—and that's all it was, a guess, armored in the certainty of the age—provided a framework into which Moore's forecasters eagerly fitted other incoming observations.

Two hours later, the Central Office issued its 8:00 A.M. national weather map for Thursday, along with a prediction that "the storm will probably continue slowly northward and its effects will be felt as far as the lower portion of the middle Atlantic coast by Friday night."

The Weather Bureau transmitted the map and its notes via the impossibly intricate web of telegraph wires that ran along every railroad right-of-way in the nation. The report caught the attention of fishermen in Long Branch, New Jersey, who cabled Washington: "Advise quick about storm unable decide about taking out nets."

Moore liked messages like this. They showed that his efforts to increase the bureau's credibility were beginning to pay off. The bureau's own scientists had always believed that one day they would be able to make accurate long-range forecasts; the most enthusiastic hoped they might even learn to make rain and quash hail. It was the public that had always questioned the bureau's competence. At last that skepticism was beginning to weaken. Many shippers, railroad agents, and cotton traders had grown as dependent on the bureau as the Galveston police had on electricity.

At 2:15 Thursday afternoon, Moore sent a reply to the Long Branch fishermen: "Not safe to leave nets in after tonight. Wind likely to increase from northeast beginning tomorrow morning."

Moore's telegram showed that the bureau was still convinced the storm was barreling north, bound ultimately for the Atlantic. The bureau had few hard facts about the storm, yet what is remarkable about its cables that day is the complete absence of doubt or qualification.

A week later, with Galveston in ruins, Cuba's Julio Jover paid a visit to Colonel Dunwoody. Emboldened by disaster, Jover sought to confront Dunwoody on the telegraph ban, but the conversation expanded to include the efficacy of hurricane prediction.

As the interview gained heat, Dunwoody grew frustrated. He told Jover, "You had better go to the Belen College Observatory and there study a work which I wrote about meteorology—see if what I said about the prediction of cyclones is not a question of divination, as a cyclone has just occurred in Galveston which no meteorologist predicted."

Jover, incredulous, paused a moment. He said, slowly, as one might address the inmate of an asylum: "That cyclone is the same one which passed over Cuba."

"No sir," Dunwoody snapped. "It cannot be; no cyclone ever can move from Florida to Galveston."

M *Is for* Missing

AT 7:00 A.M. Galveston time, Thursday, September 6, Joseph Cline made the station's morning observations, coded them, and had a messenger carry the report to the Western Union office on the Strand, where it entered the great surge of weather details that crowded the nation's telegraph lines that morning, and every morning. Joseph reported normal atmospheric pressure of 29.974 inches and a temperature of 80 degrees, markedly lower than the night before. The sky was clear and blue. Such fair weather must have been reassuring to Joseph and Isaac— the best evidence yet that the tropical storm was at that moment racing toward the Atlantic. Only much later, as meteorologists came to understand the strange physics of hurricanes, would such intervals of fair weather in the path of a tropical cyclone take on a more menacing cast.

In Washington, a legion of clerks at the Central Office processed the incoming blizzard of weather data and quickly constructed the morning's national map, which the office then telegraphed back to every station in the country. Each station then added a local and regional forecast prepared by headquarters, set the map in type, and printed copies for distribution to newspapers, post offices, boards of trade, seamen's taverns, and other public institutions.

The map that reached Erie, Pennsylvania, Thursday morning showed a vast low-pressure zone over the Pacific Coast. The base of the low stretched from Los Angeles to El Paso. From there, the low spread north to Spokane, Washington, and the Canadian border. But two high-pressure zones still held the rest of the nation's weather in check. Temperatures again soared into the nineties in Cincinnati, Davenport, Green Bay, Louisville, Washington, and Chattanooga. Even in cool green La Crosse, Wisconsin, the temperature hit 94 degrees. A brief notation on the map read: "The tropical storm has moved from Key West to Tampa, Florida."

In fact, the storm never did pass directly over southern Florida. Blocked by one of the high-pressure zones, it executed an abnormal left turn that put it on a course directly toward Galveston, eight hundred miles away across the superheated Gulf. The high pressure had caused a change in the seasonal pattern of winds sweeping off the Atlantic. Instead of blowing toward the northwest, these winds now blew mainly west, and carried the storm toward the Texas coast.

Only the storm's outer bands reached Florida. The winds in Key West, Tampa, and Jupiter did reach gale force, but caused little damage other than knocking out the fragile telegraph link between Key West and points north.

Where the Thursday-morning weather map should have displayed temperatures for Key West, the Central Office inserted only the letter *M*, for *missing*.

The Devil's Voice

WHAT THE *LOUISIANA*'s thirty passengers must have thought Wednesday as the steamship passed the striking red-and-black storm flag at Port Eads, Louisiana, is anyone's guess. For some passengers, no doubt, the prospect of a storm was an exciting one, just the thing to yield a good story to tell the friends and relatives who would meet the ship in New York the following week. Others took comfort in Captain Halsey's obvious confidence. If there was any serious threat to the ship's welfare, surely the captain would proceed no farther. A few passengers did not see the storm flag. They were seasick and already considered death an attractive option.

Once past the bar off Port Eads, the *Louisiana* accelerated. The muted booming of the ship's steamplant became an even thrum. Smoke from her stack blew *forward* over the starboard rail in a long black smudge that flurried cinder upon the sea.

Captain Halsey ordered the decks cleared and hatches sealed, but the thought of turning back did not occur to him. He held the *Louisiana* to its southeastward course throughout the night, despite the rising wind and seas.

At 6:00 A.M. Thursday he checked the ship's barometer and saw the mercury at 29.60 inches, nearly three-tenths below normal. The wind still came from the north-northeast, but at intervals circled until it came directly from the north.

The storm was a cyclone and by now Captain Halsey, veteran of so many such tropical storms, had to know it.

By ten that morning, the storm was much worse. The barometer dropped another third of an inch, to 29.25. The depth of the decline was troubling in itself, but the speed of the descent was what most captured Halsey's attention. The first decline, to 29.60 inches, had taken all night. This latest had taken four hours.

Horizontal rain clattered against the bridge with the sound of bullets against armor. Wherever the wind gained entry, it spoke. It moaned among the cabins and corridors like Marley's ghost. The hull flexed. Beams twisted. To the passengers, the ship seemed on the verge of disintegration.

At noon, Halsey ordered a sharp reduction in speed. He wanted only enough forward drive to let him maneuver and keep the ship's bow pointed into the oncoming wind and waves.

The barometer continued sinking. At one o'clock, Halsey checked the glass and saw the mercury "had fallen to the remarkable figure of 28.75." He had never seen a lower reading. He believed the *Louisiana* had arrived at the heart of the storm, for the wind now shifted wildly from one direction to another. "I do not like to speak of anything outside of the log record," Halsey said, "but I think the wind was blowing at the rate of more than 100 miles an hour."

Wave after wave washed the ship's deck and thundered against the cabin ports. By now all thirty passengers were sick beyond fear. At one point a giant wave struck the ship from behind just as it slid into a valley between two other mountains of water. In an instant, the ship was buried bow to stern under tons of green sea and foam.

The *Louisiana* rose clear, her deck like the rim of Niagara Falls. Another wave caught the ship broadside and flushed seawater down her ventilation shafts into the engine room.

It was at this point that Halsey estimated the velocity of the wind at 150 miles an hour.

•   •   •

THE TRANSFORMATION WAS stunning: One moment a nondescript tropical storm, the next, a hurricane of an intensity no American alive had ever experienced. The storm did not grow through some gradual accretion of power; it exploded forth like something escaping from a cage. The Weather Bureau of 1900 had a code word for winds of 150 miles an hour—*extreme*—but no one in the bureau seriously expected to use it.

The storm had undergone an intensification known to late-twentieth-century hurricanologists as explosive deepening, but the Weather Bureau of Isaac's time had no idea such a dramatic change could occur. As the twentieth century closed, hurricane experts still did not understand what caused it. There were theories, however. For a storm to grow so quickly, some researchers proposed, it had to encounter an additional atmospheric force—an upper-level vortex, perhaps, or a fast airstream that somehow set the storm spinning more and more rapidly. Hugh Willoughby, head of the National Oceanic and Atmospheric Administration's (NOAA's) Hurricane Research Division, proposed that explosive deepening could be caused when a storm passed over the Loop Current, a branch of the Gulf Stream that propels warm water through the Straits of Florida.

The Loop may have been in place in the summer of 1900. The Gulf was hot to begin with because of ambient high temperatures and because so far in that season there had been no other hurricanes to roil and cool the waters. The Loop brought a deep channel of warmth that the wind and rough seas could not have cooled. If present, it would have been directly in the path of the storm of 1900 when it exited Cuba. "If a storm runs over the Loop," Willoughby said, "it's got essentially an infinite source of heat."

No one knows whether crossing the Loop triggered the storm's electric growth. What is certain, however, is that for the storm to have generated winds of the velocity reported by the *Louisiana*'s Captain Halsey, it had to have formed an open, circular core of extremely low pressure.

Isaac and his peers in the Weather Bureau preferred to call this the focus, or center. They shunned the term *eye,* coined by the Spanish and used so freely by Spanish captains. It was too romantic, too anthropomorphic. In the age of scientific certainty, one could not allow one's judgment to be clouded by mere poetry.

AT THE VERY center of the eye, the air is often utterly calm. Sailors throughout history have reported seeing stars at night, blue sky during the day. Often, however, the eye is neither clear nor cloudy, but filled with a liquid light that amplifies the stillness, as if the world were suddenly fused in wax. The sea, however, is anything but calm. Freed abruptly from the wind, waves from all quadrants of the eyewall converge at the center, where they collide and compound to form sudden mountains of undirected energy.

Sunlight playing in the eyes of cyclones produced colors that drove brave seamen to their knees. Captains reported olive-green clouds and a spectral blue light that stained sails and the faces of men until all seemed turned to ice. In 1912, the Reverend J. J. Williams of Black River, Jamaica, saw the sky begin to bleed. "Around the entire horizon was a ring of blood-red fire, shading away to a brilliant amber at the zenith. The sky, in fact (it was near the hour of sunset), formed one great fiery dome of reddish light that shone through the descending rain."

The eyewall is an impossibly hostile realm where air flowing toward the center reaches its highest velocity. Observers trapped in a cyclone's eye consistently reported hearing a great roar as the calm passed and the opposite eyewall approached. The frightened Malay crew of a ship off Sumatra called this chorus the Devil's Voice. To Gilbert McQueen, commanding a ship bound for London, the eyewall sang its advance in "numberless voices, elevated to the highest tone of screaming."

One of the strangest encounters with the eye was that of Capt. William Seymour, of Cork, Ireland, and his brigantine *Judith and*

Esther, as the ship made for Jamaica in the summer of 1837. Seymour sailed into one of four hurricanes that scoured the Caribbean that summer within days of each other.

The storm knocked the ship onto her side three times, the third time just as the ship was leaving the eye. Once again the ship righted, but now something profoundly peculiar occurred that piqued great excitement among seekers of the Law of Storms. Lt. Col. William Reid wrote at once for more details.

Captain Seymour replied: "For nearly an hour we could not observe each other, or anything but merely the light; and most astonishing, every one of our finger-nails turned quite black, and remained so nearly five weeks afterwards." He could not explain it. "Whether it was from the firm grasp we had on the rigging or rails I cannot tell, but my opinion is, that the whole was caused by an electric body in the element. Every one of the crew were affected in the same way."

Such phenomena, however, were only sideshows to the most important feature of the eye, its plummeting pressure. Normal pressure at sea level is 29.92126 inches, or 14.6969 pounds per square inch. In the wall of the eye, spiraling and ascending winds lift air at over a million tons per second. As the air soars, pressure at the surface falls. Air within the eyewall rises with so much force it literally lifts the surface of the sea, one foot for each one inch of barometric decline. The lowest barometric reading ever recorded was 26.22 inches, during Hurricane Gilbert in 1988. Gilbert raised the level of the sea by over three feet.

Doctors have long been tantalized by persistent anecdotal evidence that a sudden, severe drop in atmospheric pressure can trigger premature labor and cause aneurysms to burst. Seismologists have wondered whether such a decline could rupture an already-fragile fault. Early observers of hurricanes often claimed that earthquakes acccompanied the worst storms, but William Redfield and Lieutenant Colonel Reid debunked their accounts, attributing the tremors to the interplay of thunder, wind, and imagination. One later incident, however, has resisted explanation. On September 1, 1923, a severe typhoon struck Japan, coming ashore first at Yokohama, then moving to Tokyo. As the storm raged, an intense earthquake occurred. The quake crumpled buildings and set fires; the typhoon whipped the fires into a firestorm. A Weather Bureau meteorologist, C. F. Brooks, argued that low pressure and high water, acting in concert, might have caused the earthquake. He calculated that a two-inch drop in pressure lessened the load on a single square mile of land by roughly two million tons. At the same time, a ten-foot increase in the depth of the sea caused by the wind pushing water toward shore *increased* the load by about nine million tons. The sudden differential, he argued, might have been enough to fracture a fault line already stressed to its limits.

The storm and earthquake together killed 99,330 people. Another 43,500 simply disappeared.

No one in Isaac's time would have believed such low pressures could occur. Until September 1900, any measurement under 29 inches was considered an error until proved otherwise.

IN GALVESTON, THURSDAY, Isaac Cline noted in the station's Daily Journal the presence of scattered clouds and fresh northerly winds. He noted, too, that at 2:59 P.M. 75th meridian time—1:59 Galveston time— he had received an advisory from Washington stating that the tropical storm was now "central over southern Florida." He saw no cause for concern.

That evening, he climbed to the roof of the Levy Building and recorded a temperature of 90.5, the highest temperature so far that week. The wind, he saw, was from the north at thirteen to fifteen miles per hour. The barometer read 29.818 inches, just a hair lower than the evening before. He saw scattered clouds. The bureau used a ten-point cloud scale, with ten the maximum. He rated the sky at four.

He checked to make sure all the instruments were secure. He walked down to the office, composed a coded telegram to Washington, and gave this to a messenger. Then Isaac walked home.

Squadrons of fat blue dragonflies zigzagged across his path. He nodded to friends and acquaintances, smiled at casual quips about the heat. The horses especially seemed to move more slowly.

Perhaps he felt a mixture of relief and disappointment. The tropical storm was centered over Florida—that meant soon it would cross to the Atlantic, where it would become the concern of other observers in Savannah, Charleston, and Baltimore. He was glad it was gone. Storms brought damage and extra work, and extra work was not something he needed right now.

On the other hand, storms were exciting and gave the bureau a chance to prove its worth. The sight of the red-and-black storm flag raised high over the Levy Building never failed to set Isaac's heart pounding.

No one ever remembered a nice day. But no one ever forgot the feel of a paralyzed fish, the thud of walnut-sized hail against a horse's flank, or the way a superheated wind could turn your eyes to burlap.

Swells

THE HURRICANE HAD begun sculpting the Gulf the moment it left Cuba and now it transmitted storm swells toward Galveston.

Waves form by absorbing energy from the wind. The longer the "fetch," or the expanse of sea over which the wind can blow without obstruction, the taller a wave gets. The taller it gets, the more efficiently it absorbs additional energy. Generally, its maximum height will equal half the speed of the wind. Thus a wind of 150 miles an hour can produce waves up to 75 feet tall. Other conditions, such as the chance superimposition of two or more waves, can cause waves to grow even bigger. The tallest wave on record was 112 feet, but occurred amid steady winds of only 75 miles an hour.

In a cyclonic system, the wind spirals to the left, but the waves continue forward along their original paths at speeds far faster than the storm's overall forward velocity. The forward speed of the storm of 1900 was probably no greater than ten miles an hour, but it produced swells that moved at fifty miles an hour, and began reaching the Texas coast fifteen hours after their formation.

Soon after the waves left the cyclone, they changed shape. They retained their energy, but lost much of their height and their jagged crests. They became long, easy undulations, like the grease-smooth swells that Columbus spotted on his first voyage.

As soon as they reached the Texas coast, however, they changed shape again. Whenever a deep-sea swell enters shallow water its leading edge slows. Water piles up behind it. The wave grows again. It is this effect that makes earthquake-spawned tsunamis so deceptive and so deadly. A tsunami travels across the ocean as a small hump of water but at speeds as high as five hundred miles an hour. When it reaches land, it explodes.

Heat

CAPT. J. W. SIMMONS, master of the steamship *Pensacola,* had just as little regard for weather as the *Louisiana*'s Captain Halsey. He was a veteran of eight hundred trips across the Gulf and commanded a staunch and sturdy ship, a 1,069-ton steel-hulled screw-driven steam freighter built twelve years earlier in West Hartlepool, England, and now owned by the Louisville and Nashville Railroad Company. Friday morning the ship was docked at the north end of 34th Street, in the company of scores of other ships, including the big Mallory liner *Alamo,* at 2,237 tons, and the usual large complement of British ships, which on Friday included the *Comino, Hilarius, Kendal Castle, Mexican, Norna, Red Cross, Taunton,* and the stately *Roma* in from Boston with its Captain Storms. As the *Pensacola*'s twenty-one-man crew readied the ship for its voyage to the city of Pensacola on Florida's Gulf Coast, two men came aboard as Captain Simmons's personal guests: a harbor pilot named R. T. Carroll and Galveston's Pilot Commissioner J. M. O. Menard, from one of the city's oldest families.

At 7:00 A.M., Captain Simmons ordered the crew to raise steam and make for the Bolivar Roads, the channel at the east end of Galveston Island that connected the bay to the Gulf. A left turn would have taken him toward Houston. He turned right and entered the Gulf.

The weather was clear but hot. Excessively hot, especially considering the early hour. Simmons pulled out a handkerchief and wiped the sweat from his face. By habit he checked the weather display tower at the island's east end for a storm flag. He saw nothing.

He did note, however, that the *Pensacola* was alone in the Roads.

AT 9:35 A.M. Galveston time, two and a half hours after the *Pensacola*'s departure, Willis Moore telegraphed Isaac with an order to hoist a conventional storm warning. The telegram reached Isaac at 10:30. Five minutes later, Isaac raised the flag.

The bureau's forecasters in Washington had changed their minds, and now believed the storm would not reach the Atlantic after all. They still considered it a storm of only moderate energy, but now seemed to think it was still in the Gulf, moving toward the northwest.

The Atlantic theory had been a compelling one, however—so much so that a vestige of it survived at the Galveston station well into Saturday morning, despite Isaac's experience on the beach. Shortly after nine o'clock Saturday morning, Capt. George B. Hix, master of the *Alamo,* walked to the Levy Building to inquire personally about the weather, as captains often did whenever the atmosphere seemed unsettled. Since dawn, Hix had watched the silvery shaft of mercury in his barometer get shorter and shorter.

In the weather office, an observer told him there was "no cause for uneasiness." A storm was indeed approaching, but it was only an "offspur" of a storm that had struck the Florida coast a few days earlier.

"Well, young man," Hix snorted. "It's going to be the damnedest offspur you ever saw."

Young man.

Not Isaac, surely. He was thirty-eight years old, which in 1900 qualified him as middle-aged. More likely the observer was Joseph Cline or the newly arrived John Blagden.

Regardless, it was a telling encounter. It suggests that Isaac had not told his fellow observers about his predawn trip to the beach, or at least had not revealed to them the depth of his concern. Or else he simply was not as worried as he later claimed.

Hix, however, hurried back to the wharf and readied the *Alamo* for storm.

By Friday afternoon, a few sea captains and their crews were still the only men who knew the storm's true secret—that it had grown into a monster. Some lived; some did not. In Tampa earlier, storm flags went up, but the schooner *Olive* set sail anyway for Biloxi, Mississippi. Now, she was missing. Two ships ran aground off Florida, their crews feared lost. The storm caught other ships as well—the *El Dorado* out of New Orleans, and the *Concho* and *Hyades,* both out of Galveston. Captain Halsey struggled to keep the *Louisiana* upright in waves whose backs were planed almost smooth by the intense wind.

By noon, the *Pensacola* was well into the Gulf. Captain Simmons checked his barometer and saw the mercury at 29.9. Over the next two hours, pressure fell nearly an inch. The wind reached gale force.

Captain Simmons stayed on course, the ship's bow aimed roughly toward the Mississippi Delta, where the state of Louisiana bulged into the Gulf.

Why he did not run can never be known, but it is likely his failure to do so was the product of those eight hundred previous voyages, his own ornery temperament, and the technological arrogance of the time—hell, the *Pensacola* was made of steel and weighed two million pounds.

Plus, he had an audience. At one point, in a show of bravado, Simmons called his guests to the barometer. "Menard," Simmons said. "Look at that glass. Twenty-eight point fifty-five. I have never seen it that low. You never have and will in all probability never see it again."

Simmons ordered all hatches sealed. The waves grew; the wind accelerated. Simmons gauged the wind at one hundred miles an hour.

Foam covered the sea. Spindrift blew in long luminous tentacles that seemed to reach for the bridge. Simmons stopped the engine. He ordered the anchor dropped, along with one hundred fathoms of chain cable, or six hundred feet.

When the anchor caught, the ship swung so that its bow faced head-on into the wind like a kite tethered to a child's wrist. It "labored heavily," Menard said, "rising off one tremendous sea and dropping on another, which jarred the vessel and made her tremble all over." Steel seams howled. The wild tumbling shattered crockery and lamps. Fragments slid in noisy herds back and forth across the deck. The captain's dog got seasick.

"It looked as if the good ship could not stand such a thumping," Menard recalled. "It was feared she would strain her plates or break some bolts, if the vessel did not break in two."

This two-million-pound steel-hulled screw-driven marvel of marine technology was in trouble—suddenly no better off than a square-rigged barkentine. Worse off, in some respects. Steamships could not broach-to the way the old wooden sailing ships could. If knocked on her side, the *Pensacola* would have sunk like a steel bearing. The pounding was the biggest worry. A ruptured seam, Menard guessed, would drive her to the bottom in five minutes.

Things like this were not supposed to happen. Not anymore. Whether the ship survived or not was now only a matter of luck.

Luck, and maybe a little quiet prayer.

FRIDAY NIGHT, DR. Samuel O. Young, the secretary of the Cotton Exchange, walked from his house to the beach. He lived at the corner of P½ and 25th, one block north of Isaac Cline's home, in a large two-story house mounted on brick pillars four feet high. On stormy nights, as lightning flashed, Young could see Dr. Cline standing on his second-floor balcony, keeping an eye on the weather. Dr. Cline, no doubt, could also see him.

As Young walked past the weatherman's house, he saw children outside, leaping about unmindful of the mosquitoes beginning to emerge from the gutters and the moist places left by Tuesday's thunderstorms.

His own children and his wife were at that moment in the sleeping car of a Southern Pacific train speeding toward Texas from the west, where they had spent the summer away from the heat and mosquitoes.

Ahead, Murdoch's pier blazed with light. The crests of incoming waves seemed nearly to touch the lamps suspended over the surf. There would be no nude bathing tonight—unlike *other* nights, when as many as two hundred men would gather in the waves beyond the reach of the lamps and swim frog-naked in the warm water. The thought of joining them had crossed Young's mind now and then, but he quickly put those inclinations out of his skull. He could see it now: a two-inch item in the next morning's *News* about the secretary of the Cotton Exchange tumbling naked among the waves.

The Gulf had grown angrier since Wednesday, when Young first had noticed the unusual height of the waves and the absence of any wind to explain their growth. "Thursday afternoon," he wrote, "the tide was again high and the water very rough, while the atmosphere had that peculiar hazy appearance that generally precedes a storm." Now it was Friday night. A robust wind raced past Young toward the Gulf, but did little to dispel the heat. The surf was rough, the tide unusually high, "though as a rule with a north wind the tide is low and the gulf as smooth as the bay."

To Young, this was additional evidence: "I was then confident that a cyclone was approaching us and accounted for the high tide by assuming that the storm was moving toward the northwest or against the gulf stream, thus piling up the water in the gulf."

The cyclone's exact location was anyone's guess. The Weather Bureau was no help. About all one could really tell from the bureau's advisories was that a storm of some sort did exist. The bureau had not yet acknowledged that the storm was a tropical cyclone. But it had to be, Young believed.

"For my own satisfaction, and at the request of friends, I constructed a chart, outlining roughly the origin, development and probable course of the cyclone."

He based his estimate of the storm's track on what he had seen in the Tuesday-morning weather map and on subsequent maps and advisories from the Weather Bureau's Central Office, copies of which came to the Cotton Exchange because of its obvious interest in weather. He placed the storm's origin somewhere south of Cuba, but assumed it would behave like most tropical storms—that it would travel northwest for a time "as cyclones always do," then curve toward the northeast for an exit into the Atlantic. He estimated the storm would strike the U.S. mainland somewhere near the mouth of the Mississippi.

"The error I made," he wrote, "was in placing the course too far to the east."

THAT EVENING, AT precisely 6:41 P.M. Galveston time, Joseph Cline took the necessary readings for the eight o'clock 75th meridian-time national observation.

Much of the day had been clear and hot, but now clouds filled the sky from horizon to horizon. Joseph rated the cloud cover at ten, the maximum. It was still hot, however. At 4:00 P.M. the temperature had been 90 degrees. Now, nearly three hours later, the thermometer still showed 90.

The barometer stood at 29.637, and rising. At midnight, when Joseph climbed to the Levy Building roof to take his last reading, he found the barometer had risen to 29.72.

"*Who Is Right?*"

IN HAVANA, FRIDAY afternoon, William Stockman dried his fingers on a towel that he kept beside his desk. He wound another piece of paper into his typewriter. A fan dangled from the high ceiling. The air was like a moist sweater.

He typed a page number at the top. Seventeen.

It was the last page of his reply to Col. H. H. C. Dunwoody's letter of Wednesday, September 5, in which Dunwoody had shown himself uncharacteristically perturbed by the Cubans and their rather pathetic cries of outrage over the bureau's telegraph ban. Dunwoody had written, "I think it would also be well for you to give me a copy of the statement of the mistakes which Jover made last year, and to which at one time you called my attention. . . . I may need this in defending my position."

Stockman believed he had more than fulfilled the colonel's request. In these seventeen pages he had given Dunwoody example after example of forecasts in which the Cubans had made alarming declarations that later proved baseless. No one could accuse Stockman of manipulating the record. Stockman had typed the Cuban forecasts and the corresponding U.S. advisories verbatim, with dates and times, so Dunwoody and his critics could see for themselves.

Stockman typed his last paragraph, and his closing—"*Very respectfully,*"—and pulled the page from the typewriter.

His shirt cuffs were moist. He aligned the pages of his letter in a satisfying stack. Seventeen pages. Eighteen, once he attached a chart of rainfall and wind. He tapped the bottom of the stack against the green felt blotter on his desk. Dunwoody wanted a defense. *This* was a defense.

There was nothing like a nice thick letter to make a man feel he had put in a good day's work. Out of prudence and pride, Stockman began rereading his own letter.

"Colonel:" it began.

Now, was that respectful enough? Should Stockman have written, "My dear Colonel," or the more formal "Sir," required in all correspondence with His Highness, Willis Moore?

No, he decided. "Colonel" was fine. A jot more familiar than "Sir," perhaps, but after all, he and Dunwoody were allies. Partners. Almost friends. Dunwoody had begun *his* letter "My dear Stockman."

Examples of Cuban errors comprised the bulk of the letter. The Cubans loved to dash off alarming forecasts. It seemed to Stockman that a big part of his job was simply to counter the panic their forecasts produced.

Stockman devoted half his letter to the storm that had come through Cuba earlier that week. A perfect example.

Nothing much had come of the storm, yet the Cubans had called the storm a cyclone ever since the first sighting in the final days of August. On Wednesday, September 5, Jover had actually called it a hurricane.

Jover's forecast had moved Stockman to add a few reassuring words to his own advisory: "No dangerous winds are indicated."

Any comparison of U.S. and Cuban forecasts regarding this latest storm, Stockman assured Dunwoody, "will show that the forecasts of this Bureau were verified in every particular; and that the conditions which obtained did not warrant the issuance of a forecast likely to cause any alarms whatsoever."

All in all, Stockman felt, it was an excellent letter: muscular, understated, full of detail. Eighteen pages, yes, but every word in those eighteen pages had value. Stockman sealed the letter.

It was Friday, September 7, and from the look of the latest observations telegraphed from St. Kitts, Barbados, and the other West Indies stations, the weekend would be a peaceful one. The entire season had been peaceful. No hurricanes at all, other than the imaginary ones concocted by Jover and Gangoite. Any rational man could see the need for limiting the telegraphic flow of their reports.

These people—they saw hurricanes in their sleep.

FATHER GANGOITE REMAINED troubled by atmospheric signs that suggested the storm, while no longer a threat to Cuba, had undergone a dramatic transformation.

He saw a large and persistent halo around the moon, which indicated the presence of the high, thin clouds first identified by Father Vines as signs of a hurricane. Gangoite was up at dawn the next day, composing a dispatch for *La Lucha*. He would have to deliver it by hand.

"At day-break," he wrote, "the sky was an intense red, cirrus clouds were moving from the W by N and NW by N, with a focus at these same points; these are clear indications that the storm had much more intensity and was better defined than when it crossed this island. It is, we think, central in Texas, probably at the WSW of San Antonio and northward of the city of Porfirio Diaz."

He could not resist tweaking the Americans and their mistaken belief that the storm would cross to the Atlantic, as if storms could behave one way, and one way only.

"Now some articles have been written saying that the disturbance from the SE had moved by the first quadrant out over the Atlantic; we think however that we still have it in sight as it passes through the Gulf, and that it is at present in the 4th quadrant, between Abilene and Palestine.

"Who is right?"

PART III

Spectacle

———————————

OBSERVATION

Saturday, September 8: Buford T. Morris, a real-estate agent who lived in Houston but spent weekends in Galveston at his house a few blocks from the wharf, happened to look out his bedroom window at first light.

"The sky seemed to be made of mother of pearl; gloriously pink, yet containing a fish-scale effect which reflected all the colors of the rainbow. Never had I seen such a beautiful sky."

The Pensacola

EARLY SATURDAY MORNING the *Pensacola* swung from her anchor in seas turned luminous by lightning and exploding rain. Each great swell seemed to bring the ship closer and closer to disintegration. Captain Simmons and his two guests, Menard and Carroll, held tight to rails and bulkheads, trying hard in the manner of the age not to show their fear. All night the ship's steel beams howled like wolves. Wind keened among the deck rails and boom wires. To the first officer, it seemed as if the ship were caught at the convergence of two storms, a gale from the north and a hurricane from the east, that together produced a tornado. Menard agreed. Only a confluence of storms, he believed, could produce such intensity.

Dawn brought little relief. Green swells walled the ship. At intervals visibility fell to zero. It was impossible to open one's eyes against the horizontal rain.

At 10:30 that morning, the anchor fractured. The ship's bow pivoted from the oncoming seas like a horse pulled into a sudden turn.

Captain Simmons ordered his crew to play out two hundred fathoms of nine-inch hawser from the stern, which together with the chain-cable still trailing from the bow had the effect of slowing the ship's landward drift and stabilizing its motion. The thumping stopped, but the ship now rode parallel to the oncoming crests and slid deep into the troughs between waves, a deadly place.

Simmons ordered a sounding and found the ship was in twenty fathoms of water, or 120 feet. He estimated its position at about 115 miles southeast of Galveston. The storm seemed to be shoving the *Pensacola* directly toward the city.

If Simmons was right, then Galveston lay directly in the great storm's path. It would arrive, he knew, without warning, and there was nothing he could do to sound the alarm.

Delight

AT DAWN SATURDAY two men stood on the beach, apparently out of sight of each other. One was Isaac Cline, who stood with his watch cupped in his palm, glancing from its face to the sea and back again. The other was his neighbor, Dr. Young. Both men had come to the beach for essentially the same reason.

Dr. Young watched the waves attack the streetcar trestle, which through an act of supreme confidence had been built over the Gulf itself. Waves now crashed over the rails and exploded against the pilings in vertical geysers of arctic-white spray.

Dr. Young stayed only a few moments. The sight was all the confirmation he needed. "I was certain then we were going to have a cyclone." He walked into the city, and went directly to the Western Union office on the Strand, where he composed a telegram addressed to his wife, still aboard that Southern Pacific train from the west.

It was a measure of the age that Dr. Young had such complete faith in Western Union's ability to find his wife during the train's brief stop in San Antonio.

He asked her to wait in San Antonio until he sent word for her to continue to Galveston. "I told her that a great storm was on us."

Legend holds that the sea convinced Isaac of the same thing—that he raced back to the office, galvanized the station into a flurry of action, then sped back to the beach and warned everyone he saw to flee the city or retreat to the center of town. Later Isaac took personal credit for inciting six thousand people to leave the beach and its adjacent neighbor-hoods. If not for him, he claimed later, the death toll would have been far higher. Perhaps even double.

But Isaac's response, and that of his station, was in reality more ambivalent. A few hours after Isaac's trip to the beach, the *Alamo*'s Captain Hix made his visit to the station—the visit in which he was told the coming storm was an innocuous "offspur" of one that had struck Florida. At about nine o'clock that morning, Theodore C. Bornkessell, Isaac's printer, left work to go to his cottage in the city's west end and passed the home of an acquaintance named E. F. Gerloff, who asked about the storm. Bornkessell replied there was nothing to worry about.

John Blagden, the observer assigned to Galveston on temporary duty, reported spending much of Saturday answering telephone calls from worried civilians, but it is by no means clear that he conveyed to these callers any great sense of danger. He conceded, later, "The storm was more severe than we expected."

About midmorning, Isaac himself walked to the Strand and there told several wholesale merchants that he expected minor flooding. He advised them to raise their goods three feet off the ground.

Many residents said the storm came utterly without warning. None had the slightest inkling that it might be a hurricane. One resident, Sarah Davis Hawley, noted that even as late as Saturday afternoon, despite the wind and unusually dark skies, "we weren't at all apprehensive." Another survivor, R. Wilbur Goodman, spent Saturday morning swimming and chatting with friends at the YMCA, and went home on what proved to be the last trolley of the day. The car was crowded, but "there was no talk of the storm."

Partly this was the fault of the Weather Bureau—its forecasters had failed to identify the storm as a hurricane and to recognize that it was not following the rules. The bureau's West Indies service was so busy trying to downplay the danger and show up the Cubans that it apparently missed whatever signs the Cubans saw that convinced them the storm had suddenly become more violent. And Willis Moore's obsession with control and public image guaranteed that no one in the Galveston office would even whisper the word *hurricane* without a formal authorization from Moore himself.

It was also the fault, however, of the city's newspapers and the editorial customs of the time. Certainly anyone who read that morning's Galveston *News* could be forgiven for not taking the storm too seriously.

At the turn of the century, newspaper editors expected readers to read everything and packed their pages tight with items that ranged in length from a single sentence to several full columns. They sprinkled news throughout each day's edition with what late-twentieth-century readers would consider mindless abandon. Late-breaking stories got shoehorned into whatever space happened to be available, because composers had neither the time nor the will to break apart existing plates of type. On Sunday, September 2, for example, a reporter told in extraordinary detail the story of a well-dressed young man beheaded by a switch locomotive in a freak accident on Galveston's wharf—how the head had disappeared, and no one knew the man's identity. The reporter even gave readers the color of the dead man's underwear. Later that night, at about 3:00 A.M., police found the man's head (it had been deposited atop an axle housing, hat still in place) and soon afterward identified the victim as an engineer off the steamship *Michigan* who somehow had stumbled in front of the locomotive. The editors ran both stories, four pages apart.

In fact, Saturday's edition of the *News* was a gold mine of weather information, in the sense that fragments of the story were lodged throughout the paper like nuggets on an abandoned claim. Nearly everyone in Galveston read the *News* that morning. They found the first weather story on page 2—a report about a storm that had struck the Florida coast. The second item was only one sentence long and appeared on page 3, describing how the same storm was "raging" along the Louisiana and Mississippi coasts as of 12:45 A.M. Saturday, the time at which the dispatch was filed.

On another page, the newspaper published the routine daily weather forecast out of Washington:

"For western Texas, New Mexico, Oklahoma and Indian territory: Local rains Saturday and Sunday; variable winds.

"For eastern Texas: Rain Saturday, with high northerly winds; Sunday rain, followed by clearing."

The most substantial story appeared on page 10 and reported that the Weather Bureau now believed the tropical storm in the Gulf "instead of moving north, had changed its course," and was moving toward the northwest. "The early indications were that the storm would probably strike land somewhere east of Texas, and make its way across land westwardly." The report downplayed the storm. "The weather bureau officials did not anticipate any dangerous disturbance, although they were not in a position to judge just what degree the storm may reach or develop when it strikes Texas."

Early Saturday morning, apparently just before deadline, someone at the paper added a paragraph to this story, seeking to pack the paper with the freshest news possible. "At midnight the moon was shining brightly and the sky was not as threatening as earlier in the night. The weather bureau had no late advices as to the storm's movements and it may be that the tropical disturbance has changed its course or spent its force before reaching Texas."

There was other news, of course. The Galveston *News,* like most papers of the day, gave extensive coverage to foreign events. On Saturday, the Boxer Rebellion in China dominated the front page. But the *News* also covered the most insignificant stories. It reported the newest arrivals at the Hotel Grand and the Tremont Hotel, and the general comings and goings of Galveston's citizens. Saturday's paper noted, for example, that a boy named Louis Becker had left town on Friday to attend school in

Carthage, Missouri. The Reverend W. N. Scott of the First Presbyterian Church returned on Friday from a summer away in cooler Virginia. And W. L. Norwood departed Friday night for Buffalo to attend the National Association of Undertakers and Embalmers convention set to begin on September 11. He took his wife and his young daughter along.

In just a few hours, these reports of Friday's arrivals and departures would take on an entirely different cast, and be seen instead as stories of miraculous escape and tragic bad timing.

If there were a Pulitzer for bleak irony, however, it would go to the *News* for its Saturday-morning report on one of the most important local stories of the year—the Galveston count of the 1900 U.S. census, which the newspaper had first announced on Friday. The news was excellent: Over the last decade of the nineteenth century, the city's population had increased by 29.93 percent, the highest growth rate of any southern city counted so far. "Galveston has cause to feel proud in having grown 30 percent in ten years," the *News* reported. "That is a good record to start out with on the new decade, when the prospects are bright even to surpass it."

AT THE COMPETING Galveston *Tribune,* editor Clarence Ousley spent Saturday morning writing his editorials for the Sunday edition. He looked out the window at the harsh sky. Patches of blue still showed, but mostly he saw clouds as black and low as any he had ever seen. The storm seemed a good subject for comment. Off and on that morning he had called home for reports from his family on the condition of the surf, which his wife and children could watch from the windows of the second floor. It was very exciting—storms always were—but he did not think this one would be terribly different from any other.

"There have been high waters before, when the effect was mainly discomfort and the destruction of fences," he typed. No flood could ever exceed the high-water marks already noted on landmarks around town, he argued. "Physical geographers"—mainly Commodore Matthew Fontaine Maury—"argue plausibly, with the support of experience, that the high-water records have been the maximum of possibility because the beach at Galveston slopes so gently to the ocean depths that destructive waves will be broken and their force dissipated before reaching the shore."

He struck a reassuring note: "An inundation might be wasteful and damaging, to be sure, but there is no possibility of serious loss of life."

The *Tribune* never published the editorial. The storm flooded the presses. Many decades later, Ousley's daughter Angie would describe the flooding as an event "which did much to preserve my father's reputation for editorial profundity."

CHILDREN FOUND THE storm nothing but delightful. Henry C. Cortes of Houston was eight years old when he came to Galveston on Saturday, September 8. Early that morning his father made the impulsive decision to take the family to visit Grandmother Cortes on her birthday. Henry dressed for the day in high-laced black boots, black cotton stockings with elastic black garters, white starched linen pants that ended just below the knees, a sailor-style blouse, and a stiff hat known as a straw katy. The trip took ninety minutes. Immediately after Henry and his family left the station, they got slapped by a powerful gust of wind that lifted Henry's hat off his head. It disappeared forever. When he reached his grandmother's house around lunchtime he found the yard under two and a half feet of water. "Even so," he said, "the neighboring kids were out playing in washtubs or homemade rafts."

Throughout the city, children danced in the waters, built rafts, teased pets into leaping off porches. They converged on the beach. The surf rocketing into the sky off the streetcar trestle was easily as good as a fireworks display. That morning Mrs. Charles Vidor got a call from her cousin, excitedly telling her of the marvelous sights and urging her to bring her son down for a look. The boy had the lofty name of King. Later, after he had become one of Hollywood's most important directors, King Vidor wrote a fictional account of a hurricane for *Esquire* magazine grounded on his experience in Galveston. "I remember now that it seemed as if we were in a bowl looking up toward the level of the sea. As we stood there in the sandy street, my mother and I, I wanted to take my mother's hand and hurry her away. I felt as if the sea was going to break over the edge of the bowl and come pouring down upon us."

LOUISE HOPKINS WAS just seven years old, and found double delight in Saturday morning. It had been such a hard week. School had started. Having just turned seven, she had become eligible for first grade, a prospect that had excited her no end but also gave her nightmares and made sleeping next to impossible. Not that anyone could ever sleep well with all that heat and the huge mosquitoes that blew in through the open windows in clouds as thick as dust. The first day of school had been the worst of all. "I left home, nervously holding the hand of my big sister, my brand new lunch basket and a second-hand first-grade reader in the other." But now that particular nightmare was behind her. It was Saturday. No school. The weekend. And what a weekend it was shaping up to be. There was the delicious threat of a storm. The wind was up. Best of all, the air was cool—almost chilly. It felt so good after the long, murderously hot summer. She had heard talk from her mother and from the medical students who boarded at her house of children in other places who had actually died from the heat.

Rain threatened. She raced to her closet and threw on her "Saturday" dress—the one that could become dirty without bringing down buckets of trouble. From her porch she bellowed for her best-ever friend, Martha, across the street, and soon, like magic, Martha did emerge, clothed in her own rough-time dress. Louise's mother emerged too, scowling, shushing Louise lest her shrieking wake the herd of young doctors who had just moved in upstairs for the start of the new year of medical school.

Louise did not know what to think about the doctors. There were so many of them. Sometimes they made the dining room as crowded as the train station on Sunday morning. At times she counted as many as twenty of them at the breakfast table, including medical students who came just for the meals. They talked of such strange things and always looked at you like if you did something wrong you would end up in one of those little funny-smelling jars they kept in their rooms with those mushy red-and-pink things floating around like little dead frogs only without the skin. Some days the doctors smelled just like the bottles.

Louise's father had died when she was a baby and her mother had not remarried. Louise had two brothers, John and Mason, and a sister, Lois, who was one year older than she. Their mother had added a second floor to the house, full of rental rooms so that she could earn an income at home without leaving the children. The house was perfectly located, near the University of Texas Medical School and two hospitals. Mrs. Hopkins filled the kitchen with huge sacks of green coffee, which she roasted and ground herself. She kept great drums of lard. "Our home was not only a home," Louise said, "but a living."

It was not insured.

"Martha was as glad as I to enjoy the cool windy day," Louise said. "We were not concerned the wind was stronger and the clouds darker than usual and as far as I knew neither was my mother, busy in the house as she always was."

They played in the yard for as long as the rain let them. It came in fits, and gave them fits. With each fresh squall, they leaped laughing onto the porch. When the rain stopped, they plunged back into the yard. Mud clotted their shoes. Their dresses were soaked. This was heaven.

The high curbs along the street formed a shallow canyon through which the water ran like a broad brown river, full of all kinds of interesting things. Ragged squares of wood. Boards. Trinkets. A signboard with lettering. Even an occasional snake. Toads were everywhere, climbing into the yard to escape the water.

"As we watched from the porch we were amazed and delighted to see the water from the Gulf flowing down the street. 'Good,' we thought, 'there would be no need to walk the few blocks to play at the beach, it was right at our front gate.'"

THE ENRAGED SEA drew adults by the hundreds. A great crowd gathered at the Midway, a ten-block stretch along the beach with cheap restaurants that sold beer and boiled clams, and with ramshackle stores that peddled souvenirs, candy, seashells, and stereoscopic postcards. The adults came by streetcar, hoping maybe to ride it out over the waves, but found the car had to stop well before the beach. They walked the rest of the way through pools of water. Many described the spectacle as "grand" and "beautiful." The rain struck like pebbles. The wind flayed umbrellas to their metal spines. Men and women facing the sea found their backs soaked, their fronts mostly dry. One witness reported that a few people, "with abundant foresight, appeared on the scene in bathing suits and of course were right in it from the jump."

Walter W. Davis, who had come to town on business from Scranton, Pennsylvania, was in his hotel Saturday morning about eleven o'clock when he heard people talking about how the breakers in the Gulf had become so huge they were now destroying the small shops of the Midway.

Davis did not see much of the ocean in Scranton. This he had to see for himself.

He took one of the trolleys. The trestle, he saw, ran out over the wild surf, but no cars ran on it now. The waves crashed against the rails. Big combers rolled right into the Midway itself. "The sight was grand at the time. I watched the waves wash out and break all those shell houses, theaters and lunch rooms, until I saw that the waves were coming too close for comfort."

He turned around and headed back to his hotel. It was about 12:30 now. He discovered that the streetcars had stopped running altogether. He had to walk back, at times wading through water up to his knees. The rain "felt like hail when it struck my face."

But the storm still held a powerful attraction for him. When he reached his hotel, he did not change his clothes. He had lunch in the hotel dining room, then set out for the bay side of the island.

Here too water flowed onto the city streets, but this water came from the bay. Blown by the north wind, it climbed over the piers and onto the Strand. Water raced in from the Gulf and from the bay, the former propelled by the sea, the latter by the powerful north wind. It seemed as if Galveston were a gigantic ship sinking beneath the sea.

Davis stood on a high sidewalk. The water came in so fast he could actually see it rise. It flowed below him like a spring creek, and raised translucent fins of water behind the legs of horses. Clumps of horse excrement splashed into the current and spiraled down the block. The hulls of great ships elevated by the extreme tide now towered above the warehouses of the wharf. All the ships were tightly moored, many with anchors dropped and chains reinforcing the thick rope hawsers that tied them to the piers. All seemed to have started their boilers. Smoke blew from their funnels in jagged black clouds that tore south over the Strand.

A crate drifted past. The wood paving began to float. A man fell, laughing, and let the current sweep him half a block.

Davis watched, transfixed, until he realized the water had topped the sidewalk itself and was now rippling past the soles of his shoes. It was then, he wrote in his unschooled way, "I became to be nervous."

*   *   *

DOWNTOWN, IT WAS business as usual. Women seemed to understand that something exceptional was occurring, but the men of Galveston went to great lengths to deny the strange feel of the day. They dressed as they always did, sat down to breakfast as always, drank the usual cup or two of coffee, read the morning paper, then set off for work and walked the same routes as always, the only difference being that they were forced to hold their hats against the strong northerly breeze. On the way they saw nothing out of place—provided they chose to overlook the twelve inches of water that filled every street, and the occasional boy floating past on a homemade raft. Cabs and drays moved among the avenues as if such flooding were a daily occurrence. As always, the immense fifteen-passenger bus owned by the Tremont Hotel went to the Santa Fe depot to pick up the morning's first arrivals. It would be there even when the last train from the mainland finally reached the station, despite water that by then caressed the bellies of its horses.

"My family pleaded with me to remain at home," said A. R. Wolfram, a Galveston shopkeeper, "but I was determined to go to town. I tried to reassure them and promised that at the first signs of the storm's approach, I would return home." He did go home, for lunch, but left again to return to work, "despite the tearful pleadings of my wife and children."

Ike Kempner, one of Galveston's richest men, walked into town for a meeting with two out-of-town businessmen, Joseph A. Kemp and Henry Sayles, to discuss an irrigation contract. Joseph Kemp was visibly concerned about the weather. Ike tried to reassure him. "We have had storms before," he said. "Most of our homes are built on high stilts and the water has never come up into them. Then, too, Commodore Maury, the famed oceanographer, had recently issued a statement to the effect that storms originating in the West Indies would not place Galveston in their natural paths."

The meeting continued.

. . .

JUDSON PALMER, SECRETARY of the Galveston YMCA, a centerpiece of the city's social life, also walked to work at his usual time. He and Isaac Cline knew each other. Palmer taught the adult Sunday school at the First Baptist Church, where Isaac taught the young men's class. Palmer lived at 2320 P½ Street, three blocks from Isaac's house.

On Saturday morning, Palmer's wife, Mae, occupied herself doing the baking for Sunday dinner, while their six-year-old son, Lee, played with his beloved dog, Youno.

Most days Palmer went home for lunch, but by noon the rain was gushing from the sky. Palmer decided to stay downtown.

At one o'clock, Mae called him. She told him their yard was now underwater. What's more, she had stuck her finger in the water and tasted it. It was *salt* water. She had tried calling him from the telephone in their house, but found it was no longer operating. She walked to a neighbor's place and phoned from there. Come home, she said. Please. She was starting to get scared.

Palmer stayed at work. He joked with his coworkers, the "boys," about "frightened women." Soon, though, he did leave for home, and quickly understood why his wife had sounded so anxious. This was nothing like the other storms he had experienced in Galveston. The wind was blowing at about fifty miles an hour, he guessed. Water covered every street. He caught a ride on a passing delivery wagon.

Mae fell into his arms. She did not want to stay in the house. She saw danger. They should all go downtown, she urged, and stay in the YMCA building until the storm passed. The building was strong, stronger certainly than their house. It was three stories of brick and stone.

Judson agreed. The building *would* make a safe haven—for Mae and Lee. He, however, would stay at the house and look after it during the storm.

Mae objected. He had to come. It wasn't safe to stay this close to the beach. Powerful gusts of wind punctuated her remarks. Rain slapped the broad wood shutters she had closed to protect the windows. He simply had to come.

Judson was adamant.

She looked at him, heartbroken. But she would not leave him. If he stayed, they would all stay.

LOUISA ROLLFING SHARED Mae Palmer's fear, but had the same trouble convincing her husband of the danger.

The elder August had left home at about 7:30 Saturday morning, his usual time. He walked downtown where his crew was finishing work on the Trust Building.

Louisa had not yet grown concerned about the storm. Like her children, she at first found the storm exciting, and she reveled in the coolness of the morning. Everyone seemed to be out enjoying the breeze and watching the water that flowed between the high curbs of the street. "For a while even ladies were wading in the water, thinking it was *fun*," she said. "The children had a grand time, picking up driftwood and other things that floated down the street."

After breakfast, the two oldest Rollfing children, Helen and August, went to the beach for a closer look. They returned with stories of how the surf had grown so immense it was now breaking apart the big bathhouses.

A chill moved through Louisa. She had been to the bathhouses many times. She had walked their wooden decks high above the Gulf. These were immense structures on big thick timbers. They had been there forever. No one would have dared build such things into the North Sea off the island of her childhood. But the Gulf was far more peaceful. More like a very big lake, really, than a mighty ocean.

Her children were joking. It was just the kind of big story their father would tell until his face broke in that wonderful smile.

But Helen and little August insisted it was all true. They had seen everything—big boards flying through the air, pieces of the bathhouses simply falling into the sea.

Now Louisa believed them. "Then it wasn't fun anymore."

She sent her son downtown by trolley to the Trust Building with orders to find his father and bring him home. The water, she saw, was rising quickly and soon would reach the front door. She wanted to move to the center of the city, but she wanted her husband home. She was afraid now. She wanted all the family together.

August found his father. "Mama says to come home," he said. "She wants to move."

His father laughed, and gave the boy a message.

Young August returned home. His mother watched him wade up the front walk, alone.

Louisa glared.

The boy cleared his throat, maybe scuffed his heel against the floor. "Papa says you must be crazy, he will come home for dinner."

The water continued to rise. Louisa saw neighbors begin to leave their homes.

At last her husband did arrive—"And was surprised there wasn't any dinner."

She did not kill him, but it is likely the thought crossed her mind. *Dinner*. She had not even thought about cooking.

She was furious.

He was furious.

She was being such a woman. What was there to be afraid of? This was nothing special. Some wind, some water. So what? He shouted that she should go upstairs with the children, that he was going back to town to pay his men, and would then—and only then—return to the house.

"That was more than I could stand," Louisa said. "I stamped my foot and said some terrible thing: I told him, if he didn't go immediately and get a carriage to take us away, and we in the meantime drowned, it would be *his* fault and he would never have any peace."

Which made him angrier.

August went back downtown.

"You Can't Frighten Me"

RABBI HENRY COHEN said his last good-byes to the members of his congregation and headed for home, on foot. Most days he rode his bicycle—a new "Cleveland" model—but never on the Sabbath. When he turned the corner onto Broadway, he stopped, startled by what he saw, half expecting to hear the sound of distant cannon.

Rabbi Cohen, his wife, Mollie, and their children lived about a mile from the Gulf in a comfortable gray house raised twelve feet off the ground. It had plaster walls and a long central hall, or "hog run," that cut the house in half. On the left were the bedrooms and bath, on the right the dining room, parlor, and Cohen's library, walled with books. A narrow gallery ran across the front of the house, facing Broadway.

Cohen was known throughout Galveston as a kind of psychotherapist, although the term and profession were not yet common. People of all religions and both sexes came to him to discuss troubles they felt they could disclose only to him, including problems with their sex lives. Everyone knew the rabbi and the stories that had given him near-legendary stature—the scar on his head delivered under unclear circumstances by a rifle butt during a Zulu uprising in Africa, the story of how he had barged alone into one of the city's most unsavory bordellos to rescue a young woman held captive within, throwing her over his shoulder and bolting back into the night.

He stood now on Broadway as a long line of people struggled past him toward the city. He saw whole families and noticed many carrying hampers of clothing and food and stained-glass lamps and framed photographs, like refugees from a military bombardment—except that the children all seemed delighted. And very muddy.

A lushly planted esplanade of oleander, live oak, and Mexican dagger divided Broadway, but the heavy rains of the past month and the fresh downpours of the morning had turned the esplanade into a wonderfully slippery flume of mud, through which the children stomped and slid despite stern shouts from parents on the adjacent sidewalks.

As Cohen watched, he heard fragments of the story: The sea had risen; it had destroyed the Midway; the bathhouses were about to collapse into the Gulf; the streetcar trestle was so thoroughly undermined it could not possibly stand much longer.

Cohen realized these were indeed refugees. They had left their homes for safer ground.

It was a shock. There had been floods before, but no one seemed to get terribly upset. That's why most houses, his included, were raised on posts, and why the curbs in some places were three feet high.

He took the stairs to his house three at a time and gathered up as many blankets and umbrellas as he could find, then brought them back down to the street, where he handed them out to the people who seemed most needful, the mothers with babies and toddlers, the elderly who moved so slowly against the wind.

Mollie found a bag of apples and brought it to him. He passed these out to the children, who thanked him gaily. Mud streaked their cheeks and clotted their shoes. Many were barefoot, the boys with their pants rolled to their knees. Cohen had to smile.

He was soaked. He was also shivering, a novelty for September in Galveston. He had no more umbrellas or apples, but he stayed put out of empathy for all the dislocated families, until Mollie ordered him back inside.

The power was out, Cohen saw. With the storm shutters closed, the house was as dark as night. They ate lunch by candle flare.

"We had a storm like this in '86," Mollie said, referring to the winds and rain that had reached Galveston from the last of the big Indianola hurricanes. "My father's store on Market Street was flooded," she said, casually. She noted, however, that no flood had ever reached Broadway.

With cinematic timing, a sledgehammer of wind struck the house with so much force it knocked plaster from the walls.

"It's just a little blow," Mrs. Cohen said, softly, to the children.

She swept the plaster into a small pile. The wind grew louder. Gusts came at shorter intervals, with progressively greater power. Each brought a fresh squall of plaster.

Cohen went to the front door to gauge the storm's progress, and saw that this time the water *had* reached Broadway. A shallow current raced along the street among the legs of the refugees. The water seemed to rise even as he watched.

More people crowded the street. It was a parody of the city's Mardi Gras celebration. In the stormlight everyone looked gray and worn and thoroughly miserable. The streetcars, Cohen realized, had stopped running.

When he looked outside again a few minutes later, he saw that the water now covered the first step of the stairs to his gallery. He heard his children come up behind him. He shut the door abruptly, and turned with a big smile. "Come in the parlor, Mollie," he called. "Let's have some music!"

She looked at him as if a block of plaster had just fallen on his head. She had things to do. There were lunch dishes to clear. Plaster littered the floor and plaster dust filmed the once-gleaming tops of all the tables in the house. *Music,* Henry?

Still smiling, he gave a slight nod in the direction of the children.

Mollie saw the smile; a heartbeat later she realized it did not include his eyes.

He whispered, "I don't want them to see the water rising."

She went to the piano and opened the first book she saw, a collection of Gilbert and Sullivan songs. She turned to *Patience,* one of the rabbi's favorites.

Her fingers shook.

DOWNTOWN NO ONE paid much attention to the storm. As the lunch hour approached, men set out as usual for their favorite restaurants. One of the most popular was Ritter's Café and Saloon on Mechanic Street, at the heart of the city's most vibrant commercial quarter. It was a large, high-ceilinged chamber in the ground floor of a building that also housed a second-floor printing shop with several heavy presses. The café was well known even among out-of-town businessmen, who arranged to meet customers and associates at its bright, broad tables.

Saturday morning, Stanley G. Spencer, a steamship agent who represented the Elder-Dempster and North German Lloyd lines, arranged a lunch meeting with Richard Lord, traffic manager for George H. McFadden and Brother, a cotton exporter. The two met, exchanged greetings, and took a table.

It was a pleasure to be inside in the warm, dry restaurant. Waiters in white jackets and black pants raced from table to table, bringing cocktails and towering pints of beer and huge platters of oysters and shrimp and steaks the size of bricks. The room contained a cross-section of Galveston's commercial men, including Charles Kellner, a cotton buyer from England; Henry Dreckschmidt, an agent for the Germania Life Insurance Company; and a young man named Walter M. Dailey, a clerk with Mildenberg's Wholesale Notions.

Now and then a powerful gust of wind shook the front windows with enough force to draw the attention of the diners. Each time a customer came through the front door, the wind muscled past and threatened to strip the tablecloths from under every meal. Between gusts, the diners continued talking business with a nonchalance that had to be contrived. They were aware of the storm, and knew it was getting stronger.

"Hey, Spencer!" one man shouted, from across the room. "I've just counted and there are thirteen men in this room."

Spencer laughed. Other diners joined in, glad for the relief the laughter provided. "You can't frighten me," Spencer shouted. "I'm not superstitious."

Moments later a powerful gust of wind tore off the building's roof. The "blast effect" caused by the wind's sudden entry into the enclosed space of the second floor apparently bowed the walls to the point where the beams supporting the ceiling of Ritter's slipped from their moorings. The ceiling collapsed into the dining room, amid a cascade from the second floor of desks, chairs, and the brutally heavy printing presses.

There must have been warning. A shriek of steel, perhaps, or the pistol-crack of a beam. Some men had time to dive under the big oak bar along one wall of the room.

Spencer and Lord died instantly. Three others died with them—Kellner, Dreckschmidt, and young Dailey. Five other men were badly hurt. Ritter dispatched a waiter to find a doctor.

The waiter drowned.

Word of the collapse spread quickly. No one believed it. Crowds of businessmen converged on Mechanic Street to see for themselves. Isaac came, no doubt—his office was a block and a half away. Witnesses took the story back to their offices. Messenger boys from Western Union carried the news on their rounds. Ritter's Café was gone. Men were dead.

It was the thing that at last brought fear to Galveston.

The Lost Train

ABOUT NOON ON Saturday, two trains converged on Galveston, one from the north, the other from the east.

The first train belonged to the Galveston, Houston and Henderson railroad, and had left Houston earlier that morning with the usual crowd of sightseers, businessmen, and returning residents. It arrived at the entrance to one of the three cross-bay trestles more or less on schedule, but the crossing gave its passengers a few anxious moments.

"When we crossed the bridge over Galveston Bay, going into Galveston, the water had reached an elevation equal to the bottom caps of the pile bents, or two feet below the level of the track," said A. V. Kellogg, a civil engineer.

Even in the best weather, the trestles looked fragile. In a storm, with water nearly washing over the track and gusts of wind jostling the cars, they looked deadly.

The train took it slowly. To the passengers, three miles had never seemed so long, and there was a good deal of relief when the train reached the Galveston side and clattered back onto land, although this relief was tempered by the fact that the bay was now washing over the lowlands adjacent to the railbed.

The train traveled another two miles, until a signalman stepped out of the gloom and flagged it down. Flooding had washed out a portion of the track.

Kellogg's train stood broadside to the wind. Every now and then a strong gust rammed the car with sufficient force to bounce it on its springs. Rain coursed down the windows on the north side of the train; the south windows were nearly dry and provided passengers with a perfect if rather disconcerting view of huge breakers crashing onto the none-too-distant beach.

The conductor made an announcement: The railroad had cabled to Houston for a relief train, which would arrive on an adjacent set of tracks owned by the Gulf, Colorado and Santa Fe railroad—but not for at least an hour.

It was an anxious, uncomfortable wait. The coach was hot and muggy. Passengers opened the south windows a few inches for ventilation. The rain was so loud against the train's roof and north wall that passengers had to raise their voices to speak. All the while, they watched the water rise.

By the time the relief train arrived, Kellogg said, the water was over the rails.

The new train stopped half a mile back, where the track had not yet been submerged. Kellogg's train backed up to meet it; then he and the other passengers ran across the soggy ground and climbed aboard. The relief cars, packed now with so many freshly drenched bodies, developed a climate even more tropical than that of the original train. But at least this train began to move.

Eight to ten inches of water now covered the tracks, by Kellogg's estimate. This water was not stationary, however, like the *in situ* flooding that might accompany a heavy rain.

This water raced. When it passed over the rails the turbulence caused the surface of the water to undulate like the back of a fast-moving snake. The water moved, Kellogg said, "in a westward direction at terrific speed."

The relief train eased into the water. Its crew put on heavy boots and walked ahead, testing for undermined track and shoving aside pieces of driftwood. The men looked like clam diggers probing the mud flats for dinner.

Houses soon appeared beside the tracks, but now they looked more like houseboats. Nearly all were on pilings or brick pillars, which held them well above the water, but it was clear to Kellogg that the water had gotten deeper just in the time since the relief train's arrival.

The water got so deep it flooded the firebox of the locomotive. A geyser of steam and smoke hissed into the cab, but the engineer, already soaked and windsore, pulled down his goggles and kept the train moving, feeding it the steam left in the locomotive's boiler.

The train stopped just shy of the Santa Fe Union depot, its engine a hulk of cold iron. Male passengers disembarked first and formed a human chain in the waist-deep water, and helped the children and women move through the swift current to the station platform.

Kellogg checked his watch. The time was 1:15. The wind, he guessed, was blowing at a steady thirty-five miles an hour.

He had cabled ahead to reserve a room at the Tremont Hotel downtown and steeled himself for a long, wet walk—until he saw the Tremont's horse-drawn bus waiting in front of the station, with fifteen people already seated. The water was up to the seat bottoms. He waded aboard. The bus plowed its way to the hotel.

Some of the new arrivals resolved to wait out the storm in the station, which seemed to be the sturdiest building around. The first floor was flooded, so they climbed to the second, picking their way carefully up a staircase lighted only by the "eerie" glare of a few railroad lanterns. One elderly man, believed to be some sort of scientist, carried a barometer in his baggage and now propped the device on the floor. "Every few minutes," according to one account, "he would examine it by the flickering railroad lantern and tell the people that the atmospheric pressure was still falling and that the worst was yet to come."

This did not endear him to the other passengers. Later, some would express an interest in dashing the barometer against the floor.

Another passenger from the Houston train, David Benjamin of the Fred Harvey chain of railroad eating houses, set out from the station to keep a business appointment two blocks away.

The man he had planned to meet was gone. Benjamin, perhaps thinking the storm soon would subside, made an appointment to return at three o'clock.

"It was all I could do to get back to the station," he said, "and it is needless to say that I never kept the appointment."

He was not worried about the storm, however. And no one else seemed terribly worried either. Galveston apparently took such things in stride.

The first "intimation" of the true extent of the disaster, Benjamin recalled, "came when the body of a child floated into the station."

THE SECOND TRAIN, operated by the Gulf and Interstate line, was coming from Beaumont, Texas, although many of its passengers were from New Orleans and other points in Louisiana. About noon it was rolling slowly along the flooded tracks on the Bolivar Peninsula, a slender finger of the mainland east of Galveston that was separated from the city by the ship channel. The tracks ended at Bolivar Point, near a tall lighthouse operated by keeper H. C. Claiborne and his assistant, who lived in two pretty houses on the lighthouse grounds. The train consisted of one locomotive and two coaches packed with ninety-five passengers, including John H. Poe, a member of the Louisiana State Board of Education. Poe lived in Lake Charles, Louisiana, the town where Louisa Rollfing had first experienced America. Friday night he had caught a

Southern Pacific train out of New Orleans for a business trip to Galveston. He had reached Beaumont early Saturday morning, and changed trains for the last leg of the journey.

At Bolivar Point, the train was to be run aboard a big ferry, the *Charlotte M. Allen,* for a brief voyage across the ship channel to Galveston.

Poe watched as the ferry fought its way from Galveston toward Bolivar through swells so high they broke over its bow. Black smoke from the ship's funnel rocketed south with the wind. Now and then the ship disappeared behind curtains of rain.

The captain steered the ship well to the north of the Bolivar pier to compensate for the wind, but apparently failed to gauge its true strength. He tried again and again to bring the ferry to the pier. Crewmen stationed along the ship's rails held tight against the wind and the rocking of the hull.

The captain gave up.

To Poe and his fellow passengers, accustomed to the ease and can-do precision of transportation at the turn of the century, the sight of a ferry captain giving up and turning back was astonishing. And troubling.

The train remained in place a few moments, as if stunned by this act of technological betrayal. Steam exhausted from its cylinder housings gouged the water covering the tracks. The conductor ordered the train back to Beaumont. As the engine pushed the cars slowly backward, water began flowing into the coaches.

Poe had been watching the lighthouse. Swells broke high against its base and at times cast spray nearly its full height, but it seemed the strongest thing in sight. Except for the lighthouse and the cottages of its keepers and the crown of an occasional live oak, all he saw was water. The rain sounded as if a hundred men with ball peen hammers had stationed themselves along the north side of the coach.

The train halted.

The lighthouse was a quarter mile away.

Eighty-five passengers resolved to stay with the train, believing it heavy enough to withstand the storm. A train, after all, was the biggest, strongest thing most people knew.

Poe did not trust it. He did not like the way the coach shimmied in the wind. He did not like the way the water seemed to converge from the north and south shores of the peninsula, or the speed at which it rose. Small waves now broke across the open platforms at each end of the car.

Poe and nine other passengers abandoned the train. Keeping close to one another, they moved slowly across the flooded plain toward the lighthouse. The eighty-five others remained aboard.

Scores of other storm refugees already were inside the lighthouse. They had gathered first at keeper Claiborne's house, which stood on a shallow plateau that constituted the only high ground for miles around. But the water had risen too fast. Claiborne rigged a lifeline from his house to the lighthouse door. Men held the rope with one hand, and carried women and children to the door on their backs.

By the time Poe arrived, nearly two hundred people were inside the lighthouse. The darkness of the shaft was pierced only by the gray light from the doorway and a window high up the lighthouse shaft. When he looked up through the murk, he saw two hundred people staring down from seats they had claimed along the spiral stairway that rose one hundred feet through the core of the lighthouse. He and the other train refugees were the last to enter before the sea blocked the door.

Just before he stepped inside, Poe looked back at the train. Torrents of rain obscured his view, but he thought the train had begun moving again. Smoke billowed from its stack and tumbled away over the sea.

Soon the rain and spindrift blocked his view completely. He stepped inside, wondering if he had made the right choice.

. . .

SOMEWHERE DOWN THE track, the train stopped again. Maybe the water drowned its fire, or shoved an obstacle in its path. Maybe a freak gust simply blew it from the tracks.

By Sunday morning, all eighty-five passengers were dead.

OVER THE DIN of the storm, Poe and the others heard what sounded like an artillery bombardment. They soon realized the soldiers at Fort San Jacinto on Galveston Island, just across the channel, had begun firing the fort's heavy guns. The guns boomed well into the night. Marie Berryman Lang, daughter of the assistant lighthouse keeper, remembered it all so clearly: the waves that slammed against the lighthouse as the water rose within its base and drove the two hundred refugees ever higher up its spiral shaft; the heat and desperate humidity that caused the children to cry for water; and all the while, beyond the chaos, that lonesome booming of the guns, like the drumbeat of an Army cortege.

"It was the poor soldiers," she learned the next morning, "crying for help."

A Gathering of Toads

As THE DAY progressed, Isaac Cline grew increasingly concerned about the storm. He only had to look out his office window to see that the strong north wind had pushed the waters of Galveston Bay over the wharf and into the streets of the city. By afternoon, the Gulf and the bay seemed about to converge. Clearly something extraordinary was happening—and yet there had been so little clear warning. Friday night the barometer had actually risen, and he had seen nothing of the brick-red sky that was thought to herald a hurricane. The only true sign of danger lay in the great swells, which in the few hours since his dawn visit to the beach had grown to even greater size.

Now the telephone at the station rang incessantly. He heard fear in the voices of the men and women at the other end. They told him fantastic stories about water up to their necks, waves striking their front doors, the collapse of the big bathhouses along the beach, and a strange inundation of tiny frogs—thousands of them. And he had seen the remains of Ritter's with his own eyes.

"The storm swells were increasing in magnitude and frequency and were building up a storm tide which told me as plainly as though it was a written message that great danger was approaching," he wrote later. He drove, he claimed, from one end of the beach to the other, shouting a warning to everyone he saw. "I warned the people that great danger threatened them, and advised some 6,000 persons, from the interior of the State, who were summering along the beach to go home immediately. I warned persons residing within three blocks of the beach to move to the higher portions of the city, that their houses would be undermined by the ebb and flow of the increasing storm tide and would be washed away. Summer visitors went home, and residents moved out in accordance with the advice given them. Some 6,000 lives were saved by my advice and warnings."

His story, however, does not mesh well with other accounts of the day. Of the hundreds of reminiscences in the archives of Galveston's Rosenberg Library, none mentions Isaac Cline aboard his sulky sounding the alarm. And there simply were not enough locomotives or coaches to accommodate the crush of refugees that, if his account were correct, would have sought to flee the city throughout the morning. The last train to arrive was Kellogg's GH&H train from Houston, at 1:15 P.M.; it could not have survived the journey back to the mainland. R. Wilbur Goodman took the last trolley of the day toward the beach and heard no talk of the storm among his fellow passengers. Many people did eventually leave their homes, but only after water began flowing over the wood planks of their galleries and under their front doors. By 2:30 P.M., Galveston time—the time Isaac says he recognized "that an awful disaster was upon us"—the streets within three blocks of the beach were already impassable.

Isaac's and Joseph's accounts diverged in subtle ways that seemed to shed light on their later estrangement.

Isaac reported that at 2:30 P.M. he sat down to write an urgent cable to Willis Moore, "advising him of the terrible situation, and stat[ing] that the city was fast going under water, that great loss of life must result, and stress[ing] the need for relief." He gave this to "my assistant," Joseph L. Cline, to carry to the telegraph office. "Having been on duty since 5 a.m. [four o'clock Galveston time], after giving this message to the observer, I went home to lunch."

Joseph gave himself a less passive role. "At 3:30 p.m. [2:30 Galveston time] I took a special observation to be wired to the Chief at Washington. The message indicated that the hurricane's intensity was going to be more severe than was at first anticipated. About this time, my brother paused in his warnings long enough to telephone from the beach the following fact, which I added to the message: 'Gulf rising rapidly; half the city now under water.' Had I known the whole picture, I could have altered the message at the time of its filing to read, 'Entire city under water.'"

Joseph enciphered the message, then fought his way to the Strand. "The entire pavement of wooden blocks throughout the business section was afloat and up to the level of the raised sidewalks, bobbing up and down like a carpet of corks." In places, he said, the water was knee-deep. He went first to the Western Union office, but learned its wires had been down for two hours. He walked to the nearby Postal Telegraph office, and heard the same news. "I made my way painfully back again, through the top crust of wooden blocks, to the weather bureau."

It suddenly dawned on him to use the telephone. He called the telephone company and asked for a direct long-distance connection to the Western Union office in Houston, "at the utmost speed."

The operator refused. She had four *thousand* calls ahead of his, she told him. He tried to convince her this was urgent government business. She stood her ground.

Joseph asked for the manager, Tom Powell, whom he knew. Joseph explained the situation and its urgency. But why, if Isaac had so widely sounded the alarm, did Joseph have to explain anything at all? And why did the operator refuse his request?

Powell came through. Joseph got his direct connection to Western Union in Houston. He dictated the telegram. It was truly a transitional moment: There he was, at the cusp of the twentieth century, using the telephone to send a telegram.

He told Western Union the message was to be kept absolutely confidential. "The two cities," Joseph explained, "were traditional rivals." He did not want Houston to learn yet that its arch-rival in the race for deepwater dominance now lay under the converging waters of the Gulf and bay. "I explained that the facts in the message were the property of the Weather Bureau and of the Government, and were not for public release except from Washington."

Isaac, meanwhile, was on his way home. Along the way he encountered Anthony Credo, who lived near the beach in a big two-story house with his wife and nine children.

Credo had eleven children in all, but two daughters now had families of their own and lived elsewhere. Neither was at the Credo house on Saturday. A son, William, was also absent, spending the day at the home of his fiancée.

Credo was headed for his own home, and walked part of the way alongside Isaac.

Isaac seemed worried. He told Credo he was afraid he had underestimated the storm. "Dr. Cline told Papa that this storm would be more dangerous than any of the others we had had before," said Credo's daughter Ruby. "Dr. Cline didn't like the way the water was rising; the winds from the northeast had increased in a matter of minutes."

Credo walked quickly to his house and gathered his family together. His conversation with Isaac had left him deeply troubled. He told his family to get ready to leave as quickly as possible. Then he and his wife did something that to Ruby's young eyes was positively extraordinary: They began chopping holes into the parlor floor.

Soon Isaac's route took him past the home of Judson Palmer, the YMCA secretary. Just then Palmer happened to be looking out the door to see how much higher the waters had risen.

Palmer hailed Isaac, who waded toward him. Apparently Palmer was having second thoughts about staying in the house. He asked Isaac his opinion as to the safest course—move downtown, or stay?

Stay put, Isaac said. He told Palmer his house seemed well built and sturdy and would do fine and that his family would be safer there than anywhere else. Isaac said he was on his way to his own house, and planned to stay there until the storm was over.

For Palmer, this must have been especially reassuring.

Later, with mournful clarity, Isaac wrote in his official report, "Those who lived in large strong buildings, a few blocks from the beach, one of whom was the writer of this report, thought they could weather the wind and tide."

But Isaac wasn't alone in seeing his own house as a fortress. Apparently the Cline house was considered among the staunchest in the neighborhood. "Many went to his house for safety as it was the strongest-built of any in that part of town," John Blagden said.

By the time Isaac got home, the water in his yard was waist deep. And wherever an object protruded from the water, there were toads. Tiny ones. Dozens. "Every little board, every little splinter, had about twenty or fifty toad-frogs on it," one witness remembered. "I never seen so many toad-frogs in all the days of my life."

JOSEPH LEFT FOR the house an hour or so after Isaac, and arrived about 5:30 P.M. The water was by then waist deep, Joseph said.

Neck deep, Isaac said.

Joseph was amazed to find that fifty people from the neighborhood had taken shelter in the house, including whole familes and the contractor who had built the house. "He knew better than anyone," Joseph said, "that its construction was of the finest and strongest materials, as my brother intended it to withstand the worst wind that ever blew."

Even so, Joseph did not trust it. The storm was worse than anything Galveston had ever experienced.

Evacuate, Joseph urged.

Stay, Isaac insisted.

PART IV

Cataclysm

TELEGRAM

Houston, Texas
7:37 P.M.
Sept. 9, 1900
To: Willis Moore,
Chief, U.S. Weather Bureau
Washington, D.C.

We have been absolutely unable to hear a word
from Galveston since 4 P.M. yesterday. . . .

        G. L. Vaughan,
        Manager
        Western Union, Houston

THE EAST SIDE

Louisa Rollfing

AUGUST ROLLFING FOUGHT his way back into the city. With each step the water seemed to rise higher up his legs, but that was impossible—nothing could make the sea rise so quickly. The storm was much worse than it had been on his way home. Now and then powerful gusts scraped squares of slate from nearby rooftops and launched them into the air as if they were autumn leaves. He saw whole families moving slowly toward the center of the city, everyone leaning against the wind. Broadway was a river of refugees. Suddenly Louisa's desire to escape the beach did not seem so crazy.

Rollfing walked to a livery stable, Malloy's, and there hired a driver and buggy and sent them to his address with orders to pick up Louisa and the children and take them to his mother's house in the city's West End. He believed it a far safer neighborhood, perhaps because it was many blocks from the ocean beaches at the east and south edges of the city. Apparently he did not take into account the fact that the bay was only ten blocks north of his mother's home. The wind was still blowing from the north over the long fetch of Galveston Bay, and with each increase in velocity drove more water into the city. Rollfing went to his shop.

At one o'clock, the buggy pulled up in front of the family's house at 18th and Avenue O½. Louisa was overjoyed. She raced through the house collecting shoes and a change of clothing for everyone, and packed these in a large hamper, but once the driver and her children and she had all climbed aboard, she realized there simply was no room left. She had to leave the hamper behind.

She held Atlanta Anna in her arms. The driver set off for the West End, no doubt first driving north toward the slightly higher ground at the center of the city, then due west. "It was a terrible trip," Louisa said. "We could only go slowly for the electric wires were down everywhere, which made it dangerous. . . . The rain was icy cold and hurt our faces like glass splinters, and little 'Lanta' cried all along the way. I pressed her little face hard against my breast, so she would not be hurt so badly. August and Helen didn't cry, they never said a word."

The driver dodged other storm refugees and great masses of floating wreckage. Judging by the quantity, whole houses must have come apart. The sky was so dark, it looked as if dusk had arrived half a day early.

"We got as far as 40th Street and Ave. H, just one block from Grandma," Louisa said. "The water was so high, we just sat in it, the horse was up to his neck in water."

The driver turned onto 40th Street. Someone shouted for the buggy to stop. "Don't go! You can't go through." The water was too deep, the caller said—there was a large hole ahead filled with water.

The driver turned the buggy around, and asked Louisa, "Where shall I go now?"

Louisa, in a nearly submerged carriage with three young children, was at a loss. "I don't know," she said.

It came to her then: August's sister, Julia, and her husband, Jim, lived in a house at 36th and Broadway, just six blocks back toward the city. The driver gently eased his horse back along Avenue H, against the flow of water and refugees.

When Julia saw Louisa and the children, drenched and windblown, she was shocked. "My God, Louisa, what is the matter?"

Clearly Julia knew nothing of the damage along the beach. Louisa quickly described conditions in the East End and how the West End too was underwater. She paid the Mallory's driver a dollar and made him promise to tell his boss, Mr. Mallory, their new destination so that Mallory could pass the message on to August.

"I was so confident that August would go there," she said, "but he didn't."

At about two o'clock Galveston time, in the midst of Louisa's drive, the wind shifted. Until then the wind had blown consistently from the north, the weaker left flank of the hurricane. Now the wind circled to the northeast and gained intensity. Isaac noticed the change, but most people, including Louisa, did not. They were too busy seeking shelter or had battened themselves within their homes. The stories Louisa told her hosts of what she had seen on her journey frightened them. With Louisa's help, they began bracing the windows and doors. They nailed an ironing board across a window. A neighbor came over with her two children seeking shelter or company and brought the total number of people in the house to ten. They closed all the upstairs doors and gathered on the stairway. They had a pitcher of water and a lantern. Soon they heard the shattering of windows and blinds in the bedrooms behind the doors they had just closed. "It sounded," Louisa said, "as if the rooms were filled with a thousand little devils, shrieking and whistling."

She watched quietly as Julia and Jim's piano slid from one downstairs wall to another, then back.

AVENUE P½

Parents and Their Choices

Sam Young

AT TWO O'CLOCK in the afternoon, Dr. Young started back to his house at the northeast corner of Bath and P½, one block north of Isaac Cline's house and adjacent to the Bath Avenue Public School. Thinking his family safe in San Antonio, he prepared for the storm's arrival—prepared, that is, to enjoy it, and savor every destructive impulse. Young was a member of that class of mostly landlocked men who believed God put storms on earth expressly for their entertainment.

Young's yard was a plateau of land five feet above sea level, yet by the time he got home he found the yard under two feet of water. This did not trouble him, for he had seen high water before. He took a chair on his gallery and watched the storm. The water rose gradually and soon began to climb the stairs toward him. Even this did not worry him. "My house, a large two-story frame building, stood on brick pillars about four feet high," he said, "so I had no fear of the water coming into my house."

A young black boy worked for Young as a valet. Young sent him home, then began closing shutters and windows and securing doors, intent mainly on getting these tasks done before nightfall.

Around four o'clock, he began to see that he had been wrong about the water. Two feet now covered his ground floor, and the level was still rising—not gradually, anymore, but rapidly. Visibly. Like water flowing into a bathtub.

Young had noticed the change in wind direction. "The wind had hauled further to the east and was blowing at a terrific rate." The shift accounted for the more-rapid increase in depth, Young knew. Galveston sat astride a portion of Texas coastline canted forty-five degrees toward the northeast. All morning the north wind had impeded the progress of the incoming storm tide, causing water literally to pile up in the Gulf. Now, with the wind blowing from the northeast, a portion of that pent-up tide—but by no means the bulk of it—began to come ashore. The wind blowing southwest along the Texas coast pushed the sea into Galveston's East Side.

More fascinated than appalled, Young moved a chair to a second-floor window and watched the water as it flowed along Avenue P½. (He makes no mention of seeing Isaac Cline or Joseph struggling home, although the last block of their journeys would have been within his view.) The water moved fastest at the center of the street where the high curbs channeled the water and vastly increased its velocity, just like the narrow pipes used by city water systems to increase water pressure. The street had become a causeway for wreckage. Young saw boxes, barrels, carriages, cisterns, outhouses, and small shacks. He watched one barrel hold its course all the way down the street. "The flow," he saw, "was almost exactly east to west."

What he did not realize, apparently, was that the flow now included corpses.

Most likely he had stationed himself at a window that faced west or south; otherwise he would have been aware of the great damage now being done to the beach neighborhoods behind his house, where immense breakers slid over the surface of the tide and broke against second-floor windows.

It was getting dark. He found a candle, lit it, then thought better—he might need the candle later on. He blew it out. There was nothing to do, he decided, but wait out the storm. He still felt unafraid. "I found a comfortable armchair and made myself as comfortable as possible." He was very glad, however, that his family soon would be snug and dry in the train station at San Antonio. "Being entirely alone, with no responsibility on me, I felt satisfied and very complacent, for I was fool enough not to be the least afraid of wind or water."

FOR OTHER FATHERS in homes not far from his the afternoon was playing out in rather different fashion. Suddenly the prospect of watching their children die became very real.

Whom did you save? Did you seek to save one child, or try to save all, at the risk ultimately of saving none? Did you save a daughter or a son? The youngest or your firstborn? Did you save that sun-kissed child who gave you delight every morning, or the benighted adolescent who made your day a torment—save *him,* because every piece of you screamed to save the sweet one?

And if you saved none, what then?

How did you go on?

Mrs. Hopkins

As LOUISE HOPKINS and her friend Martha played in the yard, they saw more strange things come floating past in the street. There were boxes and boards and bits of clothing, and now children's toys. Martha went home, fearing that soon the water would be too deep to cross, and indeed soon afterward the level rose to where water flowed into Louise's yard and into her mother's treasured garden. The sight of all that brown water destroying her mother's lovely flowers brought Louise the day's first sadness.

When Louise went inside, she saw for the first time that her mother was worried about the storm. Mrs. Hopkins was moving her great trove of cooking supplies to the second floor—her sacks of sugar, coffee, and flour (one of the most popular brands was Tidal Wave Flour). Between trips Mrs. Hopkins went to the window to watch for Louise's two brothers, who that morning had ridden their bicycles to their jobs. "She knew now with the water rising it would be impossible for them to come home the same way," Louise recalled.

Louise noticed that her kitten, a Maltese, was behaving strangely. The little thing "was restless and kept following me. I believe he was more aware of the approaching danger than I."

Her brother John arrived safely, and quickly went to work helping his mother transport the supplies to the second floor. Louise helped with the smaller things. Her brother Mason, fourteen years old, was still not home.

Once all the big sacks had been hauled upstairs, Louise's mother found the family ax and did something that just about took Louise's breath away forever. Her mother had always been so careful about the house. The house was everything. A home, an income. She kept it spotless, and polished and dusted the floors until they gleamed like the beacon of the Bolivar Light, and if you tracked mud onto these floors you knew you would not see the sunlight for the rest of your living days.

Right there, no warning, her mother lifted the ax over her shoulder and slammed it into the floor. She kept chopping until the holes were big enough to see through.

"I was amazed to see how fast the water came in under the front door and through the holes my mother had cut in the floor," Louise said. "How quickly the house was filling with water, and how difficult it was for my mother to keep her head out of the water as she reached down into the lower cabinets for the last of the groceries to be taken upstairs."

Louise looked out an upstairs window and saw that water now covered the porch rail of the house next door. Until then, she had felt mostly excitement. The morning had been full of wonders: water racing down the street, toads all over the place, her mother chopping holes in the floor, and water even inside the house. But there was something about the water so deep around her neighbor's house that took all the excitement away. "I thought of all the things that had to be left behind, and I was sad and afraid."

Her mother watched for Mason.

AT PRECISELY 2:30 P.M. Galveston time a gust of wind lifted the Weather Bureau's rain gauge from the roof of the Levy Building and carried it off toward the southwest. It had captured 1.27 inches of rain.

At 5:15, the wind destroyed the bureau's anemometer. By then the instrument had registered a maximum velocity of one hundred miles an hour.

The wind continued to intensify.

A FIGURE APPROACHED the Hopkins home, moving against the current. The water was up to his underarms. He dodged pieces of lumber and boxes and telegraph poles. Now and then a square of slate smacked the water around him. Softer things bumped against his legs, then moved on with the current.

When Mason arrived, a lightness came over the Hopkins household. It was as if the house itself had been holding its breath awaiting his arrival. He was bruised and soaked, but smiling, and Mrs. Hopkins hugged him as she had never hugged anyone before. The storm raged and water burbled up through the holes in the floor and slid in a sheet under the front door, but everyone was home and the unspoken fear that had gripped the place was suddenly gone. "We had a warm feeling of all of us being together, safely, we believed. We went upstairs in the main part of our house . . . to wait out the storm."

ALL OVER GALVESTON, there was a need for light. A craving. People needed light for themselves to ease their fears, but they also needed others to know they were still in their homes and alive. Throughout Galveston, lamps bloomed in a thousand second-floor windows. We're here, they said. Come for us. Please.

The same idea came to Louise's mother. She did not want to use a lamp, however. The house was shaking too badly. She feared the lamp would fall and set fire to the house, and then all would indeed be lost.

She dragged one of the big drums of lard to the center of the room. She found a carnival flag attached to a stick, and laid the stick across the top of the drum. She saturated a strip of cloth with lard, then draped this over the stick, one end in the lard, for a wick. "When it was lighted it gave off a dim and eerie light," Louise said. "We sat and watched it flickering and listened to the banging and howling of the storm outside."

It was oddly comfortable in the room. Almost cozy.

Until her sister, Lois, screamed.

Judson Palmer

IN THE BLOCKS behind Dr. Young's house, several families began moving toward the home of Judson Palmer, at 2320 P½. Anyone could see it was one of the strongest houses around. Isaac Cline himself had gauged it a perfectly safe haven against the storm. A neighbor couple, Mr. and Mrs. Boecker, arrived with their two children. Garry Burnett followed with his wife and his two children. Soon afterward another Burnett, George, arrived with his wife, child, and mother. The last couple to arrive was an unidentified black man and his wife who asked if they too might come inside until the storm passed. Palmer now counted seventeen people in his house, including his own wife and his son, Lee. The boy's dog, Youno, scampered wildly around the house, clearly delighted with the attention of so many big and small human beings.

At 6:00 P.M., Palmer and the other men rolled up the first-floor carpets and hauled them upstairs. They carried the furniture up next, an effort that caused them all to break a heavy sweat. With all the doors and windows shut and so many moist people inside, the house felt hot, humid, and stale. Once all the furniture was moved, everyone went to the second floor, which had four bedrooms and a large bathroom equipped with a tugboat-sized tub and a shower.

If a train had crossed the ceiling it could not have made more noise. With most of the slate shingles gone, the rain struck bare wood. Driven by the wind, it penetrated deep into the plaster. It grew cysts in the wallpaper, which popped like firecrackers. At 7:00 P.M., a gust of wind blew out the front door and its frame. The blast effect caused everyone's ears to pop.

Palmer estimated the water in the yard to be seven feet deep; in the parlor, two feet. He was downstairs monitoring its progress when the big plate-glass window at the front of the house exploded, along with its frame.

Palmer lit a kerosene lamp and placed it near the window of the front-most bedroom. The window shattered; the blinds disintegrated. Everyone retreated to the back of the house. Palmer brought the lamp. Here too the windows shattered. A chunk of plaster fell from the ceiling and crushed the lamp. Palmer closed a pair of big sliding doors. He suggested prayers and hymns. His son said, "I cannot pray," then reconsidered. "Dear Jesus," he said, "make the waters recede and give us a pleasant day tomorrow to play, and save my little dog Youno and save Claire Ousley."

Rain poured into the room. More plaster fell.

GARRY BURNETT RECOMMENDED everyone squeeze into the bathroom, arguing it was the strongest, safest place in the house. George Burnett believed no room would be safe if the house collapsed into the sea. He crawled out the bathroom window onto an upended roof that had floated against the house, and persuaded his mother, wife, and child to follow. They sailed off into the storm. The Palmers joined Garry Burnett in the bathroom. The Boecker family stayed behind in the bedroom. What the black couple did is unknown.

The water rose high onto the second floor. Gusts of wind moving at speeds possibly as great as 150 miles an hour—perhaps much higher—penetrated deep into the house. Palmer held tight to his son and braced his back against the bathroom door. His wife, Mae, hugged his neck with all her strength.

Beams fractured. Glass broke. Lumber ricocheted among the walls of the hallway outside the bath. The front half of the house tore loose. The Boeckers stood in their bedroom holding each other close as the wind peeled the house away. The bedroom disintegrated.

The water rose. The Palmers climbed onto the lip of the bathtub. Judson clamped his left hand to the shower rod and held Lee circled in his right arm. Youno was gone. Mrs. Palmer grabbed the rod with her right hand, and with her left held on to her husband and son.

The house trembled, and eased off its elevated foundation. It settled in deeper water. The water was up to Palmer's neck. He fought to keep Lee's head clear.

And Lee asked, "Papa, are we safe?"

Judson could not even see his son, for the darkness. He felt the boy's small hands holding tight. His hands were cold. Maybe Judson did have time to offer his son some reassuring lie; more likely he could not speak for the great heave of sorrow that welled up within him after his son's question. He drew his son close, but could not draw him close enough.

The roof stood up and fell upon the family. They went under the water together. Palmer came up alone. He had swallowed a great volume of water. He coughed, vomited. He saw nothing of Lee or Mae. There was no light, only motion. He could not think. His mind dimmed, came back.

And he was outside, free of the house. Treading water. He felt what seemed to be ground beneath his feet but could not get purchase. A wave threw him onto a mass of floating wreckage. Window shutters— many of them, all tied together. Someone else's raft, but it was empty now.

He called for his son and wife.

25TH AND Q

Isaac Cline

THEY ARGUED. JOSEPH wanted everyone to leave at once and head for the center of town. Isaac had faith in his house, but also argued that conditions outside had grown too dangerous, certainly for his wife, who was pregnant and ill in bed. "At this time . . . the roofs of houses and timbers were flying through the streets as though they were paper," Isaac said, "and it appeared suicidal to attempt a journey through the flying timbers." Water now covered the first floor of his home to a depth of eight inches.

At 6:30 P.M., Isaac, ever the observer, walked to the front door to take a look outside. He opened his door upon a fantastic landscape. Where once there had been streets neatly lined with houses there was open sea, punctured here and there by telegraph poles, second stories, and rooftops. He saw no waves, however. The sea was strangely flat, its surface blown smooth by the wind. The Neville house across the way now looked so odd. It had been a lovely house: three stories sided in an intricate pattern of fish-scale shingles and shiplap boards and painted four different colors. Now only the top two-thirds protruded from the water. Every slate had been stripped from its roof.

The fact he saw no waves was ominous, although he did not know it. Behind his house, closer to the beach, the sea had erected an escarpment of wreckage three stories tall and several miles long. It contained homes and parts of homes and rooftops that floated like the hulls of dismasted ships; it carried landaus, buggies, pianos, privies, red-plush portieres, prisms, photographs, wicker seat-bottoms, and of course corpses, hundreds of them. Perhaps thousands. It was so tall, so massive that it acted as a kind of seawall and absorbed the direct impact of the breakers lumbering off the Gulf. The waves shoved the ridge forward, toward the north and west. It moved slowly, but with irresistible momentum, and wherever it passed, it scraped the city clean of all structures and all life. If not for the wind, Isaac would have heard it coming as a horrendous blend of screams and exploding wood. It shoved before it immense sections of the streetcar trestle that once had snaked over the Gulf.

Something else caught Isaac's attention, as it did the attention of nearly every other soul in Galveston.

"I was standing at my front door, which was partly open, watching the water, which was flowing with great rapidity from east to west," he said. Suddenly the level of the water rose four feet in just four seconds. This was not a wave, but the sea itself. "The sudden rise of 4 feet brought it above my waist before I could change my position."

For those inside the house, it was a moment of profound terror. (Joseph claims to have been utterly calm. He says the rise occurred just after he had called his brother outside to try to persuade him, privately, that the best course was to evacuate for the center of town.) Four feet was taller than most of the children in the house. Throughout the city, parents rushed to their children. They lifted them from the water and propped them on tables, dressers, and pianos. People in single-story homes had nowhere to go. In Isaac's house, everyone hurried to the second floor. The brothers herded the refugees into a bedroom on the windward side, reasoning that if the house fell over, they would all be on top, not crushed underneath.

Isaac judged the depth of the water by its position in his house. His yard, he knew, was 5.2 feet above sea level. The water was ten feet above the ground. That meant the tide was now 15.2 feet deep in his neighborhood—and still rising. "These observations," he noted later, for the benefit of skeptics, "were carefully taken and represent to within a few tenths of a foot the true conditions." It was, he acknowledged, incredible. "No one ever dreamed that the water would reach the height observed in the present case."

ONE BLOCK NORTH, Dr. Young observed the same impossible increase. Since five o'clock he had noted a change in the direction of the wind. It had begun circling to the east and gained velocity, as did the current. "The debris fairly flew past, so rapid had the tide become," he said. At 5:40 P.M., he observed a sudden acceleration of the wind. He knew the time exactly because his clock had stopped and he had just finished resetting it by his watch. (Clocks began to stop throughout Galveston, as wind burst into homes and buffeted the pendulums that drove them.) He looked out a west window at a fence he had been using to gauge the depth of the water. "And while I was looking, I saw the tide suddenly rise fully four feet at one bound."

Moments later, he saw houses on the south side of P½ between 25th and 26th—half a block north of Isaac's house—collapse into the water, among them the pretty one-story home of a man named Alexander Coddou, the father of five children whose wife happened to be off the island. The houses fell gracefully at first. One witness, watching the same thing happen in his neighborhood, said houses fell into the Gulf "as gently as a mother would lay her infant in the cradle." It was when the current caught them and swept them away that the violence occurred, with bedrooms erupting in a tumult of flying glass and wood, rooftops soaring through the air like monstrous kites.

Dr. Cline's house, Young saw, was still standing, although floating debris had torn away his first and second-floor galleries.

SOON THE WATER on Isaac's first floor was over nine feet deep. The wind tore at the house like an immense crowbar. The ridge of debris came closer and closer, destroying homes south and east of Isaac's house and casting them against his walls. Isaac's house rocked and trembled, but remained firmly footed on its pilings. Isaac at this point still believed the house strong enough to survive the assault. He did not know, however, that the ridge of debris was now pushing before it a segment of streetcar trestle a quarter-mile long, consisting of tons of cross-ties and timbers held together by rails.

Joseph knew nothing of this either. He believed the house would fail simply because the storm was too powerful.

"Strangely enough," Joseph wrote, "amid the seething turmoil, I did not feel unduly excited. In fact, I was almost calm. I was convinced that, in some way or another, I should come out of it alive. I kept thinking of an uncle of ours, who, alone of all those aboard a sinking ship, saved himself by getting on a plank when the vessel went under and [by] drifting upon this frail support five miles to shore."

Joseph may have been calm, but he was not helping anyone else achieve such peace. "Again, as strongly as I could, I warned my relatives and friends that the house was about to collapse."

Imagine it, the atmosphere in this house. Fifty terrified men, women, and children packed into one room, Isaac's wife in bed, his three daughters petrified but snuggling close to their mother for comfort. The room is insufferably hot and moist. The walls drip condensation. Now and then rain spits through the ceiling; a pocket in the wallpaper explodes. Beside the bed stands Dr. Isaac Monroe Cline, thirty-eight years old, bearded, confident the house can endure anything mere nature can muster, but even more certain that to venture outside would be like stepping in front of a locomotive. Nearby, perhaps at the other side of the bed, stands Joseph, the earnest younger brother, apprentice-for-life, who has always always always resented Isaac's insufferable pose—that *he,* not Joseph, was the man who knew weather, *he* knew when the rain would fall, *he* knew when true danger loomed. The conversation starts quietly but soon, partly because their tempers rise, partly just to be heard over the wind, rain, and barrage of debris, they start shouting. "Are you deaf, Isaac?" Joseph perhaps cries. "What do you think that is, for God's sake? An evening breeze? This house will not stand. Out there at least we have a chance."

Isaac prevailed. Joseph, frustrated, began offering advice for how best to survive the coming collapse. "I urged them, if possible, to get on top of the drift and float upon it when the dangerous moment came. As the peril became greater, so did the crowd's excitement. Most of them began to sing; some of them were weeping, even wailing; while, again, others knelt in panic-stricken prayer. Many of them were scrambling aimlessly about, seeking what, in their fright, appeared to be vantage points."

The battering continued. By now all four galleries had been torn from Isaac's house, all slate stripped from its roof.

The trestle was a yard away.

IN DALLAS, THREE hundred miles north, the telegraph operator at the Dallas *News,* sister to the Galveston *News,* realized the steady flow of cables from the Galveston paper had ceased. The two newspapers maintained a leased telegraph line that ran directly between their editorial offices. The telegrapher at the Dallas paper keyed off an inquiry, but got no response. He tried again. Again nothing. He then tried raising Galveston over public lines by relay through Beaumont, and next by sending a message to Vera Cruz, Mexico, for relay to Galveston via the Mexican Cable Company (whose Galveston agent had only a few hours to live).

Again he failed.

At that moment, City Editor William O'Leary was in the office of the Dallas paper's manager, G. B. Dealey, showing Dealey a passage in Matthew Fontaine Maury's best-selling *Physical Geography of the Sea* that seemed to show "that destruction of Galveston by tropical storm could not happen."

The wires remained dead.

Vital Signs

SATURDAY EVENING, JOHN Blagden, the new man temporarily assigned to Galveston, found himself alone in the office. He had been in the city all of two weeks and here he was alone in the dark, facing a storm whose intensity seemed to place it in the realm of the supernatural.

The Levy Building was four stories tall and made of brick but in some gusts, Blagden said, it "rocked frightfully." Bornkessell, the station's printer, had left for home first thing in the morning. Isaac had gone home next, followed by Joseph. Ernest Kuhnel, a clerk, was supposed to be in the office but had fled the building in terror.

The storm flag was gone, as were the anemometer, rain gauge, and sunshine recorder. The telephone had stopped ringing. There was nothing for Blagden to do but watch the barometer and try to keep himself sane. He estimated the wind at 110 miles an hour.

The hurricane had set a course toward Galveston soon after leaving Cuba, and had stayed on that course ever since, as if it had chosen Galveston as its target. It had a different target, however. The great low-pressure zone that had formed over the Pacific Coast earlier in the week had progressed to where it now covered a broad slice of the nation from Texas to Canada. The hurricane saw this low-pressure zone as a giant open door through which it could at last begin its northward journey.

The storm's track intersected Galveston's coastline at a ninety-degree angle, with the eye passing about forty miles west of the city somewhere between Galveston and the Brazos River. Meteorologists discovered this later when officers aboard an Army tug stationed at the mouth of the Brazos reported a pattern of winds that showed the eye had passed somewhere east of their position. The pattern in Galveston indicated the eye had passed to the west of the city. This was the worst-possible angle of approach, for it brought the hurricane's most-powerful right flank directly into the city.

Blagden knew nothing of the storm's track. What he did know was that the first shift in wind direction, from north to northeast, had brought a sudden acceleration in wind speed. And now he sensed the wind beginning to shift again, toward the east. Impossibly, the change seemed to bring another increase in velocity. Gusts struck the building like cannonballs.

Barometric pressure had fallen all day, but at five o'clock Galveston time it began to fall as if someone had punched a leak into the instrument's mercury basin. At five, the barometer read 29.05 inches.

Nineteen minutes later, 28.95.

At 6:40 P.M., 28.73 inches.

Eight minutes later, 28.70.

An hour later, the barometer read 28.53 inches, and continued falling. It bottomed at 28.48.

Blagden had never seen it that low. Few people had. At the time, it was the lowest reading ever recorded by a station of the U.S. Weather Bureau.

In fact, the storm drove the pressure even lower, although just how far will always be a mystery. The bureau's instruments in the Levy Building captured pressures well away from the center of the eye, where the pressure would have been lowest.

Barometers elsewhere in the city got widely varied readings. In Galveston harbor, the first mate of the English steamer *Comino,* moored at Pier 14, recorded in the ship's log a pressure of 28.30 inches, and noted: "Wind blowing terrific, and steamer bombarded with large pieces of timber, shells, and all manner of flying debris from the surrounding buildings." At one point the wind picked up a board measuring four feet by six inches and hurled it with such velocity it pierced the *Comino*'s hull. The hull was built of iron plates one inch thick. In the train station, the scientist with the barometer—apparently unaware of his fast-eroding popularity—called out a pressure of 27.50 inches, and announced that against such impossibly low pressures "nothing could endure."

Years later, scientists with NOAA put the lowest pressure of the storm a notch lower, at 27.49.

In 1900, however, even Blagden's reading of 28.48 stretched credibility. "Assuming that the reading of the barometer reported at Galveston the evening of the 8th was approximately correct," wrote one of Moore's professors, carefully hedging for error, "the hurricane at that point was of almost unparalleled severity."

The highest speed recorded by the Galveston station's anemometer before it blew away was 100 miles per hour. The bureau later estimated that between 5:15 P.M. and 7 P.M. Galveston time, the wind reached a sustained velocity of "at least" 120 miles per hour.

Most likely the true velocity was far greater, especially within the eyewall itself. Gusts of two hundred miles an hour may have raked Galveston. Each would generate pressure of 152 pounds per square foot, or more than sixty thousand pounds against a house wall. Thirty tons.

As John Blagden sat in his office, powerful bursts of wind tore off the fourth floor of a nearby building, the Moody Bank at the Strand and 22nd, as neatly as if it had been sliced off with a delicatessen meat shaver. Captain Storms of the *Roma* had practically bolted his ship to its pier, but the wind tore the ship loose and sent it on a wild journey through Galveston's harbor, during which it destroyed all three railroad causeways over the bay. The wind hurtled grown men across streets and knocked horses onto their sides as if they were targets in a shooting gallery. Slate shingles became whirling scimitars that eviscerated men and horses. Decapitations occurred. Long splinters of wood pierced limbs and eyes. One man tied his shoes to his head as a kind of helmet, then struggled home. The wind threw bricks with such force they traveled parallel to the ground. A survivor identified only as Charlie saw bricks blown from the Tremont Hotel "like they were little feathers."

All this was nothing, however, compared to what the wind had been doing in the Gulf of Mexico. Ever since leaving Cuba, the storm had piled water along its leading edge, producing a dome of water that twentieth-century meteorologists would call a storm surge.

Early scientists believed that reduced pressure alone accounted for storm tides. By the mid-nineteenth century, however, they came to understand that a one-inch decline in pressure raised the sea only a foot. Thus even a pressure as low as 27.49 inches would cause the sea to rise only two and a half feet. Yet the Galveston storm shoved before it a surge that was over fifteen feet deep.

The single most important force needed to build a storm surge is wind. A strong wind will develop a surge in any body of water. A fan blowing across a water-filled container will cause the water to swell at the downwind side. Strong winds blowing over some of Minnesota's biggest northern lakes will pile ice to the height of a McDonald's sign. One of the deadliest storm surges in American history occurred on Lake Okeechobee in Florida, in 1928, when hurricane winds blowing across the long fetch of the lake raised a storm surge that killed 1,835 people.

Another ingredient is geography. In 1876 Henry Blanford, a meteorologist in India, proposed that the configuration of the Bay of Bengal contributed greatly to the immense storm tides that came ashore during typhoons. Blanford thought of these tides as great waves. Every cyclone raised them, "but it is only when the wave thus formed reaches a low coast, with a shallow shelving foreshore, such as are the coasts of Bengal and Orissa, that, like the tidal wave, it is retarded and piled up to a height which enables it to inundate the flats of the maritime belt, over which it sweeps with an irresistible onset."

Despite such reports, Isaac and his colleagues in the bureau believed that a hurricane's most lethal weapon was the wind. They did not see the parallels. Isaac, like the famous Commodore Maury, believed the shallow slope of the seabed off Galveston would wear down incoming seas before they struck the city, and had argued in his 1891 *News* article that mainland areas north of Galveston Bay would serve as basins to capture whatever floodwaters a storm did manage to drive ashore.

The hurricane of 1900 would cause a hasty reevaluation. In October, in the Weather Bureau's *Monthly Weather Review,* one of the bureau's leading lights, Prof. E. B. Garriott, belatedly observed that Galveston's geography and topography in fact "render it, in the presence of severe storms, peculiarly subject to inundation."

A storm's trajectory can also increase the destructive power of a surge. If a hurricane strikes at an oblique angle, it spreads its storm surge over a broader swath of coast, thereby dissipating the surge's depth and energy. The Galveston hurricane struck the Texas coast head-on, at a nearly perfect ninety-degree angle, after traveling a long, unobstructed fetch of some eight hundred miles. The track focused the onshore flow directly into the city.

The track produced another lethal effect, however. It brought north winds to Galveston Bay twenty-four hours before landfall. Throughout most of Saturday, these intensified to gale force and finally to hurricane force. Due north of Galveston Island, the bay offers an unobstructed fetch of about thirty-five miles (about the same fetch as presented by Lake Okeechobee). And just as in the freak Galveston blizzard of February 1899, the wind blew the water out of Galveston Bay—this time into the city itself.

In effect, the storm's trajectory made Galveston the victim of two storm surges, the first from the bay, the second from the Gulf, and ensured moreover that the Gulf portion would be exceptionally severe. Throughout the morning, the north winds kept the leading edge of the Gulf surge out at sea, banking the water and transforming the Gulf into a compressed spring, ready to leap forward the moment the winds shifted.

The first shift, from north to east, began at about two o'clock Saturday afternoon, Galveston time. This allowed some of the Gulf surge to come ashore. Water flowed over the Bolivar Peninsula and began rising within the shaft of the Bolivar Light. It flowed too over Fort San Jacinto and Galveston's East Side, where it met the floodwater already driven into the city from the bay. The reason so many men and women in Galveston began furiously chopping holes in their beloved parlor floors was to admit the water and, they hoped, anchor their homes in place.

At 7:30 P.M., the wind shifted again, this time from east to south. And again it accelerated. It moved through the city like a mailman delivering dynamite. Sustained winds must have reached 150 miles an hour, gusts perhaps 200 or more.

The sea followed.

Galveston became Atlantis.

The Wind and Dr. Young

ABOUT SEVEN O'CLOCK, Dr. Young heard a heavy thumping that seemed to come from a downstairs bedroom on the east side of his house. He lit the candle that he had held in reserve and walked toward the hall stairwell, the candle throwing only a shallow arc of light on the floor around him. Pistol-shot drafts penetrated deep within the house and caused the candle's flame to twist, but did nothing to cool the rooms. At the Levy Building about then John Blagden was recording a temperature of 84.2 degrees. The shock of each thump vibrated through the floor of Young's house. It was as if someone were standing in the downstairs bedroom striking the ceiling with a railroad mallet.

The stairwell appeared ahead as a large black rectangle stamped from the floor, and the closer Young got, the deeper the candlelight traveled. It should have shown him stairs and the wood slats of the banister, but he saw neither, only an orange glow undulating on the opposite wall like sunlight off a floating mirror.

Water, he realized. The sea had risen within his house nearly to the top step. The heavy thudding from the bedroom had to be furniture. A bureau, perhaps, bumping against the ceiling as the water rose and fell.

Young set the candle on the floor and walked to the door that led to his second-floor gallery. He opened it. "In a second I was blown back into the hall."

The wind snuffed the flame, then blew the candle and its holder to the far reaches of the house. From within the darkness of the hall, the doorway appeared as a rectangle of wild gray air. The power of the wind shocked Young; it also inflamed his curiosity. Another man might have sought shelter in one of the second-floor bedrooms, but Young, drawn by the sheer power of the storm, fought his way back toward the door.

He kept close to the wall. He winched himself forward from door-knob to doorknob. At the door, he fastened his hands around the frame and hauled himself outside. "The scene," he said, "was the grandest I ever witnessed."

It was as if he were aboard a ship in a storm. Waves swept through his neighborhood. One witness said the waves looked like the "sides of huge elephants." Each embodied a destructive power nearly beyond measure. A single cubic yard of water weighs about fifteen hundred pounds. A wave fifty feet long and ten feet high has a static weight of over eighty thousand pounds. Moving at thirty miles an hour, it gener-ates forward momentum of over two million pounds, so much force, in fact, that at this point during the storm the incoming swells had begun destroying the brand-new artillery emplacements at Fort Crockett, which had been designed to withstand Spanish bombardment. Debris made the waves especially dangerous. Each wave propelled huge pieces of wreckage that did to houses what the reinforced prow of Captain Nemo's *Nautilus* did to great warships. One man reported dodging a giant piano embedded in the crest of a wave, "its white keys gleaming even in the darkness."

The only other house still standing belonged to a family named Youens, with the mother, father, son, and daughter still inside. Two minutes later, Young saw the Youens house begin a slow pirouette. "It turned partly around and then seemed to hang as if suspended."

At about the same time, the wind changed direction from east to southeast, and again intensified. Young felt himself compressed against the wall of his gallery. "Mr. Youens' house rose like a huge steamboat, was swept back and suddenly disappeared," Young said. He thought of the family inside. "My feelings were indescribable as I saw them go."

Now he was alone, his house an atoll in a typhoon. The water continued to rise. "At one bound it reached my second story and poured in my door, which was exactly thirty-three feet above the street. The wind again increased. It did not come in gusts, but was more like the steady downpour of Niagara than anything I can think of."

The wind tore loose one of the posts that supported the gallery roof. The post struck Young, gashed his head, and left him dazed, but he did not fall. The wind held him in place. The door seemed about to tear loose. If the house fell, he resolved, he would grab the door, rip it free, and use it as a raft.

Slats from the gallery rail blew away "like straws." The remaining posts cartwheeled into the sea. The gallery roof lifted upward as if hinged, then blew away over the top of the house. With a shriek of wood and iron the gallery floor wrenched away and barged west.

Young remained pinned to the wall, one foot inside the doorway. He could not move. "It was an easy thing to stay there for the wind held me as firmly as if I had been screwed to the house."

The wind grew even stronger. Young estimated it reached 125 miles an hour. "The wind at 125 miles an hour is something awful," he said. "I could neither hear nor see."

He turned his head against the rain until he was looking inside the house. The rain slammed against the interior walls with such force it exploded in pixels of light. "The drops of rain became luminous," he said. It looked "like a display of miniature fireworks."

The wind grew so strong it planed the sea. "The surface of the water was almost flat. The wind beat it down so that there was not even a suspicion of a wave."

He could not open his eyes. A lion roared at his ears. That his house still stood seemed impossible. "I began to think my house would never go."

He gripped the facing of the door. He waited. He planned to kick his raft free of the house at the first sign of collapse. He did not have long to wait.

ALL OVER GALVESTON freakish things occurred. Slate fractured skulls and removed limbs. Venomous snakes spiraled upward into trees occupied by people. A rocket of timber killed a horse in midgallop.

At the expensive Lucas Terrace apartment building, Edward Quayle of Liverpool, England, who had arrived in Galveston with his wife three days earlier, happened to walk past a window just as the room underwent a catastrophic depressurization. The window exploded outward into the storm along with Mr. Quayle, who rocketed to his death trailing a slipstream of screams from his wife.

At another address, Mrs. William Henry Heideman, eight months pregnant, saw her house collapse and apparently kill her husband and three-year-old son. She climbed onto a floating roof. When the roof collided with something else, the shock sent her sliding down into a floating trunk, which then sailed right to the upper windows of the city's Ursuline convent. The sisters hauled her inside, dressed her in warm clothes, and put her to bed in one of the convent cells. She went into labor. Meanwhile, a man stranded in a tree in the convent courtyard heard the cry of a small child and plucked him from the current. A heartbeat later, he saw that the child was his own nephew—Mrs. Heideman's three-year-old son.

Mrs. Heideman had her baby. She was reunited with her son. She never saw her husband again.

THE HOUSE SHUDDERED, shifted, became buoyant. For a few queasy moments, Dr. Young felt himself exempt from gravity's effect. The time had come. He tore the gallery door loose and dove for the sea. Like the survivor of a sinking liner, he kicked hard to put distance between himself and the house. "The house rose out of the water several feet, was caught by the wind and whisped away like a railway train, and I was left in perfect security, free from all floating timber or debris, to follow more slowly."

The current drew him over the city. He saw few landmarks but believed he soon passed over the Garten Verein. Moments later he too careened toward the Ursuline convent, but his door got caught in a large whirlpool of water and wreckage. "I was carried round and round until I lost my bearings completely."

When the whirlpool dissipated, the inflowing sea again captured his raft. It swept him northwest for fifteen blocks until his door docked itself against a mound of wreckage. "It was very dark, but I could see the tops of some houses barely above the water; could see others totally wrecked, and others half submerged." He saw no lights, however. And no people. "I concluded that the whole of that part of town had been destroyed and that I was the only survivor."

He remained in that place aboard his door for the next eight hours. The wind rippled over his clothing. Porcupine rain jabbed his scalp and hands. Blood seeped from the gash in his head. In all that time he heard only one human voice, that of a woman somewhere in the distance crying for help. He had never been so cold in his life.

What Joseph Saw

SOMETHING STRUCK THE house with terrific force. The house moved. It slid from its foundation and began to list. Joseph was standing near a window beside Isaac's oldest children, Allie May and Rosemary. "As the house capsized, I seized the hand of each of my brother's two children, turned my back toward the window, and, lunging from my heels, smashed through the glass and the wooden storm shutters, still gripping the hands of the two youngsters. The momentum hurled us all through the window as the building, with seeming deliberation, settled far over. It rocked a bit and then rose fairly level on the surface of the flood."

Joseph and the two girls found themselves on top of an outside wall. They saw no one else. "All the other occupants of that room, nearly fifty men, women and children, it appeared, were still trapped inside, for the house had not yet broken up."

The only exit from the house was the now-horizontal window through which Joseph and the children had passed. Joseph lowered the top half of his body through the window and shouted, "Come here! Come here!"

No one came. No one called out. The space below the window was utterly black. Periodically the house rose with the current, then settled, raising the water within to the level where the window glass had been. Anyone still inside would be completely submerged.

Joseph had heard that drowning men seized anything that came near. He sat on the window casing and began swinging his feet in the water. "I had hoped that some of the trapped ones within the room might catch my feet and so be pulled out," he said. "My efforts were wasted and I finally gave them up. I have no words to tell the agony of heart I experienced in that moment."

Ruby Credo

As soon as Ruby Credo's parents finished chopping holes into the floor of their parlor, they began preparations to evacuate to higher ground. If Dr. Cline planned to ride out the storm in his own house, that was his choice. Anthony Credo had no intention of doing likewise. He and his family were just about to leave when a neighbor, Mrs. Theodore Goldman, appeared at the door with her son, hoping to shelter in the Credos' house. Mrs. Goldman did not trust her own house, she said. Her husband did, however, and he was still there. He refused to leave.

The Credos put on some coffee and gave Mrs. Goldman and her son some dry clothes. In that short time, the water deepened to the point where Credo saw that leaving would be more dangerous than staying.

He had built a storm shelter behind his house, a one-room chamber atop six-foot posts. He believed, at first, that his children would be safest there. He swam them over one by one. As he watched other houses in the neighborhood disintegrate, he changed his mind. He retrieved his children. If something terrible happened, he wanted his family together. His two grown daughters were with their husbands, and he presumed them safe. His son William, visiting his fiancée, was a grown man and could take care of himself. It was the young ones he worried about most—little Ruby and her sisters, and son Raymond. The shuttling back and forth to the storm shelter unnerved him. He could carry only one child at a time.

"The water was rising rapidly to the second floor," Ruby said, "so Papa helped us climb from the outside through dormer windows to the attic bedrooms, where Mr. Goldman and his mother had moved. The water had risen so fast Mama hadn't time to grab her cherished black satin corset from downstairs." The family had little to do but watch the storm intensify. "We stood at the windows and watched the houses around us break up, wash away, and become battering rams to knock and tear others apart as they were hurled and swept about. The water kept rising; the sounds of the storm were frightening; the house creaked and groaned as if it were in some kind of agony."

Night had fallen. Ruby sat on the corner of a bed opposite Mrs. Goldman and her son. The wind accelerated. A streetcar rail pierced the roof and penetrated the floor between Ruby and the Goldmans. No one was hurt.

The house began to move. The wind lifted the roof, then dropped it. Falling wreckage pinned Ruby's mother, but Anthony Credo managed to pull her free. She bled heavily from head lacerations. Credo tore strips of cloth from her clothes to make bandages.

All this occurred in darkness.

The house eased from its foundation, slid through a shallow westward arc, then began to float. Credo gathered his family and ordered everyone out the dormer windows. The Goldmans declined to leave.

"When our house left the ground, we grabbed at anything washing by, as Papa had instructed us to do, but it was all you could do to stay on a piece of wood," Ruby said.

Waves broke upon the family and scattered them. Credo herded them together again. The cycle repeated itself.

In darkness.

The sea pushed the family north, everyone alive, everyone more or less intact, although Ruby's mother looked like a soldier wounded in the Spanish-American War.

They drifted. Credo shouted orders. Between waves, he kicked himself up from the water as high as he could, to count his family and keep anyone from straying. One wave drove a telegraph pole into the back of Raymond's head. It knocked him out and dug a severe gash in his scalp. Even in the darkness, Anthony Credo could tell the fluid pouring off his son's head was blood. Credo held Raymond with one arm and kept himself afloat with his other, struggling to hold Raymond's head out of the water and still keep track of the rest of his family.

Credo was tired. He believed his son dead, or nearly so. Several times he considered letting Raymond go. Mrs. Credo would not let him. She was not ready. She still had hope.

The storm was more intense than ever, but for a time the Credos saw a full moon behind thin clouds. An inverted roof floated past. Credo ordered everyone aboard. One daughter, Florence, helped him pull Raymond into the roof. Credo went back into the water. He did not want to risk tipping the raft. Mrs. Credo held Raymond close.

At first the roof proved an effective lifeboat, but soon it began to break apart. Credo watched for something better. An upended porch floated near. It looked sturdier than the roof. Credo shouted for everyone to abandon the roof and climb onto the porch.

Ruby's elder sisters Queeny, Vivian, and Ethel sat together, holding tight to one another's clothing. The porch was so stable, some of the children fell asleep. "We could lie back on these sections," Ruby said. "They were well-made, with no jagged nails or splinters to gash our bodies as we were tossed about."

Everyone relaxed. Raymond still did not move, but there was hope, now. The family was together. They would find Raymond a doctor. Everything would be all right. "We floated this way for an hour," Ruby said. "Then a piece of timber blown up by a wave struck my three sisters a terrific blow, knocking Vivian into the water and under heavy debris."

Vivian did not surface. The porch sailed on. The moon disappeared and lightning flared, the first lightning anyone could recall seeing. Big barrels of thunder rolled among the waves, and made the night even more terrifying. To Ruby, the rain was a particular torment. It "felt like bullets."

Ruby's sister Pearl was sitting peacefully upon the raft when a jagged spike of wood blew through her arm, just below her elbow. She screamed. Her mother held Pearl tight as Anthony Credo pulled the spike from her arm. Pearl writhed in utter agony. Credo applied pressure until the bleeding slowed, then bandaged it as best he could.

The porch beached itself against a reef of debris twelve feet high, near an intact house. Ruby and her family picked their way over the wreckage and climbed inside. Anthony Credo carried Raymond on his back.

Credo tallied the family's casualties: Vivian dead; Raymond clearly dying; Pearl hurt and now at grave risk of infection, fever, amputation, even death.

An unbearable list, but in fact it understated the true extent of the family's loss.

25TH AND Q

What Isaac Did

WHEN THE TRESTLE struck, Isaac was at the center of the room with his wife and his six-year-old daughter, Esther Bellew. His baby. A wall came toward him. It propelled him backward into a large chimney. There was motion. He could not see it, but felt it all around. Things fell from the sky. Furniture, books, lanterns, beams, planks. People. Children. He entered the water. Something huge caught him and drove him to the bottom. Timbers held him. He opened his eyes. He felt the water but saw nothing. It was quiet. He could not move. He knew he would die. There was peace in this. It gave him time to think. He appraised things. The only course was to welcome the sea into his body. He did so. He disappeared.

He awoke to lions. Rain came like shrapnel. He was afloat, his chest caught between two large timbers. He coughed water. He sensed burden. There was something he had to do. It was like waking to a child's cry in the night. He sensed absence.

It came to him abruptly that he was now alone.

THE BEACH

A Light in the Window

THE SCREAM HAD been shocking enough. What Louise Hopkins saw next caused her heart to leap halfway from her body.

Her sister, Lois, red-faced from the great energy she stuffed into that scream, pointed furiously at the place where the east wall joined the ceiling. At first Louise did not understand, but as she watched, she saw the wall begin to breathe. With each gust of wind, the wall moved out from the house until Louise could see the sky; then the wall wheezed back into position. There was a moon outside. Louise saw clouds rushing by overhead.

Louise looked at her mother. Mrs. Hopkins alone seemed not to be surprised. Apparently, she had been watching all along, but had not wanted to frighten her children any more than they already were.

It was time to leave, Mrs. Hopkins resolved. The house across the way, owned by the Dau family, looked sturdy, and there was a light inside. They would go there. Mrs. Hopkins worked out a plan. They would use a mattress as a raft. The Hopkins boys, both strong swimmers, would pull it across the street with Mrs. Hopkins, Lois, and Louise aboard. Mrs. Hopkins pulled sheets from the bed and tore them into strips, which she tied around her waist and the waists of her girls.

They assembled behind the big double front door, poised to exit. Every time the east wall and ceiling parted, Mrs. Hopkins would cry, "Let's go now."

But in the next instant the ceiling would settle, and Lois would shout, "Wait."

They could not muster the courage to cross. Water flowed wildly down the street. Bursts of spindrift erupted from the surface as missiles of slate and timber hissed back to earth.

The light across the way was irresistible. It offered safety, comfort, and company. "It doesn't seem so now," Louise said, "but there was such a consolation to know that somebody was still alive."

But this light, this beacon of comfort, began to move. They saw it dance from room to room. It moved toward the front door. They saw Mr. Dau carry the lantern out his front door and down his steps.

Leaving—the man was *leaving*. Like a ship captain ignoring a lifeboat adrift.

To Louise and her family it was as if hope itself had just departed.

THREE MILES DOWN the beach, the big St. Mary's Orphanage with ninety-three children inside was under siege. It was a fortress of brick and stone that rose straight out of the grass just north of the tide line, a lonely Gibraltar shrouded most evenings in blue mist. Now waves crashed against its second story. Anyone watching from outside would have seen the lights of candles and lanterns move from room to room toward the back of the orphanage as the frontmost portions of the building collapsed into the sea like icebergs calved from a glacier.

The ten sisters who ran the place herded all ninety-three children into the chapel. Sister M. Camillus Tracy, thirty-one years old, the mother superior, ordered the other sisters to tie lengths of clothesline to the youngest children, then tie one end around their waists. They formed chains of six to eight children each, roped together like miniature climbing parties. A few older children, among them Will Murney, Albert Campbell, and Francis Bulnavic remained free. Sister Camillus led the children in hymns, including the children's favorite, "Queen of the Waves."

The water rose. The children felt the concussion of each breaker as it struck the front of the building.

The sisters drew the children to the girls' dormitory at the back of the building, away from the beach. They heard the crash of wood and brick behind them as the boys' dormitory fell into the Gulf. The storm advanced through the building quickly and systematically, as if hunting the children. The chapel disappeared. Windows shattered. Hallways rose and fell like drawbridges. The children sang.

The sea and wind burst into the dormitory. In seconds, the building failed. Ninety children and all ten sisters died. Only Will, Albert, and Francis survived, all by catching hold of the same floating tree.

Later, a rescuer found one toddler's corpse on the beach. He tried lifting the child. A length of clothesline leaped from the sand, then tightened. He pulled the line. Another child emerged. The line continued into the sand. He uncovered eight children and a nun.

Sister Camillus had hoped the clothesline would save the children, but it was the clothesline, rescuers saw, that caused so many to die, tangling them in submerged wreckage.

AUGUST ROLLFING SAT alone in his shop on 24th Street waiting for his men to come for their pay. As the storm worsened, his anxiety increased. Water began coming into his shop. The wind accelerated. It rolled up the tin roofs across the way, then hurled them to the ground like spent shell casings. Boards and glass shrapneled the street. August had money for eighteen workers. No one came.

He locked his shop and set out to join his family, with absolute faith that the driver from Malloy's Livery had in fact done as he had asked and that now his family was safe at his mother's house. He struggled west. He got as far as the city waterworks at 30th Street between Avenue G and Avenue H, when the wind picked him up "like a piece of paper" and blew him out of the water onto a sidewalk. He hugged a telephone pole. In a lull between gusts, he crawled to the waterworks building and entered. He found the lobby full of storm refugees.

The building seemed sturdy enough. What worried the occupants was its tall smokestack, which swayed through the sky like a giant black cobra. If it fell—when it fell—everyone in the building stood an excellent chance of being crushed. Whenever the wind paused, a group of refugees would dash out onto the sidewalk.

Rollfing left, accompanied by two black men. They went first to a grocery store, which soon became too dangerous. They moved next to a private house. A beam fell and killed a man. They moved on, until they saw a light in the window of another store.

August and his companions banged on the door. At first, the occupants refused to let them in. Finally they relented.

It was nearly dark now. In the shuddery glow of lanterns and candles, August saw that the store was crowded with about eighty men, women, and children, all standing on countertops to keep out of the water. But the water was rising fast. August found a place on a counter. Soon the water was at his ankles, then his chest. August lifted someone else's son onto his shoulders as the water rose to his own neck.

He spent hours this way, until a man shouted, "The water is going down! Look at the door!"

The water had indeed reversed flow. The store owner pulled out a large jug of whiskey and passed it around the room. Men and women alike took great swallows and passed it on.

August wanted desperately to leave for his mother's house to join his wife and children and make sure they were still safe. The water receded quickly, but to him its exit seemed to take forever. Rain continued cascading from the darkness; the wind seemed little changed.

At last the water level fell low enough to enable him to leave. Outside, he saw that houses had been shattered and upended. He stumbled through deep holes gouged by the current, and over all manner of submerged debris. He dodged showers of timber and slate. It was dark, no lights anywhere. He fell, got up, fell again. The damage got progressively worse. Whole blocks had been crushed, others swept clean. He knew he was heading west—probably along Avenue H—but the darkness and devastation had eliminated all landmarks.

At intervals the moon emerged. How the moon could shine amid such wind and rain he did not know, but there it was, visible through a thin layer of cloud. A full moon, no less. It gave him light; it also gave him fear, for it showed him how vast the plane of devastation truly was. Spiky dunes of wreckage blocked his path. From the top of each, he saw that only a few homes still stood. To the south was a strange black shadow two and three stories high that stretched for miles like a mountain range freshly jabbed through the earth's crust.

At three o'clock Sunday morning he came to his mother's neighborhood. Only her house looked whole. Everything else had been destroyed, upended, or transported toward the bay. Relief poured into his heart. He burst into the house but found only his mother.

"Where are Louisa and the children? I don't see them."

The question surprised his mother. "August, I don't know," she said. "They are not here." When she realized that August *expected* them to be there, she too became afraid. "When did they go," she asked, "and how?"

He told her about the buggy he had sent at one o'clock and the instructions he had given the driver.

"Nobody could come here at one o'clock," his mother said. August started toward the door. "Wait," she pleaded. "Wait until daylight."

August made his way to his sister's house. He saw corpses. The short journey—only half a dozen blocks—took an hour. The sight made him half-crazy with dread. The house stood at a forty-five degree angle.

Where Julia's kitchen had been, there was now only a jagged black hole. Every shutter had been splintered, every window broken.

But there seemed to be a light within. He pounded on the front door. The door opened. He saw Julia and her husband. He saw Louisa. He saw Helen, August, and little Lanta. "Thank God," he said.

And fainted on the stairs.

25TH AND Q

Isaac's Voyage

HE WAS ALONE in the water. His family was gone. He flailed his arms and reached deep underwater and kicked his legs to feel for soft things, clothing, someone alive. He felt only square shapes, planks, serrated edges. He had been inside the house; now he was outside in darkness, in wind so fast it planed the water smooth. There was lightning. He saw debris everywhere, jutting from the sea. He saw a child. He shimmied free of the timbers and swam hard. The rain stung; he could hold his eyes open only a few seconds at a time. He came to her and felt his arm grow from the water and circle her and knew immediately the child was his Esther, his six-year-old. His baby. He spoke into her ear. She cried and grabbed him hard and put him under, but he was delighted. She asked for her mother. He had no answer. The house began to break up. He swam her away.

He was elated; he was distraught. He had found one daughter but lost everyone else. His memory of them would be tinted the yellow of lamplight. He tried to place them in the room, and by doing so, to place them in the sea. His wife had been with him in the center of the room with Esther. His two eldest daughters had been near the window, beside Joseph. Why had they not surfaced too?

Isaac and his baby drifted. There was more lightning. He coughed water through his nose and mouth. In the next flare, he saw three figures hanging tight to floating wreckage. Isaac swam Esther toward them against the wind.

He heard a shout.

Joseph Cline said: "My heart suddenly leaped with uncontrollable joy. In two figures that clung to the drift about one hundred feet to leeward, I discovered my brother and his youngest child."

Isaac: "We placed the children in front of us, turned our backs to the winds and held planks, taken from the floating wreckage, to our backs to distribute and lighten the blows which the wind driven debris was showering upon us continually."

Joseph: "Our little group now numbered five. We remained close together, climbing and crawling from one piece of wreckage to another, with each of the latter in turn sinking under our weight. At one time it seemed as though we were indeed lost. A weather-battered hulk that had once been a house came bearing down upon us, one side upreared at an angle of about forty-five degrees, at a height from six to eight feet higher than our drift. I was conscious of being direly frightened, but I retained sufficient presence of mind to leap as the monster reached us, and to get a grip with my hands on the highest edge of the wreck. My weight was enough to drag it perceptibly lower in the water, and I called my brother, who added his weight to my own."

Isaac: "Sometimes the blows of debris were so strong that we would be knocked several feet into the surging waters, when we would fight our way back to the children and continue the struggle to survive."

Joseph: "At one point, two other castaways, a man and a woman, joined us on the wreckage that, at that time, was serving us as a lifeboat. The strangers remained with us for some little time, until the man crawled up to where I sat, pulled the two children away, and tried to shelter himself behind my body. I pushed him indignantly away and drew the children back. He repeated the unspeakable performance. This time I drew out a knife that I carried, and threatened him with it."

. . .

THEY DRIFTED FOR hours aboard a large raft of wreckage, first traveling well out to sea, then, when the wind shifted to come from the southeast and south, back into the city. For the first time they heard cries for help, these coming from a large two-story house directly in their path. Their raft bulldozed the house into the sea. The cries stopped.

A rocket of timber struck Isaac and knocked him down, but only dazed him. Joseph saw a small girl struggling in the sea and assumed that somehow Esther had fallen from Isaac's grasp. He plucked her from the water and gathered her close to the other girls. Allie May, the eldest, cried out, "Papa! Papa! Uncle Joe is neglecting Rosemary and me for this strange child!"

Stunned, Joseph took a close look at the girl. It was not Esther at all. He looked over his shoulder and saw Isaac bent over his baby, shielding her from the flying debris. This girl was a stranger.

Their raft ran aground at 28th and Avenue P, four blocks from where they once had lived. They saw a house with a light in the window, and climbed inside. Safe—although one daughter had injuries that Joseph considered life-threatening.

A miracle had occurred, Isaac knew. Nothing else could explain why he and his three daughters were still alive. Yet the enormity of what he did lose now came home to him. His children wept for their mother, but soon, out of sheer exhaustion, they fell asleep. Isaac lay awake for a time, hoping his wife somehow had survived, but knowing heart-deep that she had not.

She had been very close, as it happens. Later it would seem to Isaac as if she had been watching over her family during the entire voyage, guiding them in their passage through the night until they were safely back home.

AND THERE WAS this: In the midst of the Clines' voyage, a beautiful retriever climbed aboard their raft. It was Joseph's dog. Somehow in the storm it had sensed them and swum after them. The dog was delighted to see Joseph and Isaac and the children, but sensed too that someone was missing. He went one by one to each of them, as if marking a checklist. One scent was absent. The dog raced to the edge of the raft and peered into the water. Joseph called him back. The dog stood scrabbling at the edge, obviously torn by conflicting needs. But it was clear where his passion lay. The dog ignored Joseph and prepared to jump. Joseph lunged for him, but the dog entered the sea, and soon he too was gone.

PART V

Strange News

———————

TELEGRAM

Houston, Texas

11:25 P.M.

Sept. 9, 1900

To: Willis Moore,

Chief, U.S. Weather Bureau

First news from Galveston just received by train which could get no closer to the bay shore than six miles, where Prairie was strewn with debris and dead bodies. About two hundred corpses counted from train. Large Steamship stranded two miles inland. Nothing could be seen of Galveston. Loss of life and property undoubtedly most appalling. Weather clear and bright here with gentle southeast wind.

G. L. Vaughan

Manager,

Western Union, Houston

First Glimpse

THE *PENSACOLA* DRIFTED in the old seas of the storm throughout Saturday and Saturday night. About dawn, the remains of the anchor caught something in the seabed, and again the ship swung, again her beams and plates began to bind and flex. But the barometer showed a steady rise in pressure. The storm had passed.

Captain Simmons ordered the crew to haul in the stern hawser and the anchor chain-cable, and to restart the engines. He ordered another sounding and found the ship in only eighty feet of water. Given the slope of the seabed, he estimated through dead reckoning that Galveston was now about fifty-five miles to the northwest. The ship had drifted over fifty miles. He set a course back to the city.

About noon on Sunday, Simmons spotted the coast and followed it west, looking for landmarks, but found his view blocked by squalls.

In the afternoon, the clouds began to break and the sea to gleam a rich royal blue. Simmons spotted the Galveston grain elevator, and turned toward it, but as the ship entered the Bolivar channel he and his guests fell silent.

They entered a changed world. Nothing was as it had been when the ship left. "We found a line of breakers where the jetties were, but everything on them washed away, beacons, bay lights, lightship, buoys here and there out of position," Menard said. "We discovered steamers ashore, the forts and barracks, torpedo casemate all gone, and as we entered we began to see the terrible destruction to the city, and we knew not what news to expect when we landed of our loved ones at home."

Where buildings had stood they saw great mounds of timber. Whole neighborhoods seemed to have disappeared, and the immense bath-houses were simply gone. Now and then a peculiar scent drifted to the ship from the city, and some aboard recognized it immediately as the odor of putrefaction. But to smell it at this distance—what did that mean?

No one worried much about the loss of physical property, Menard said, "but our anxiety about the loss of life was terrible."

It was about five o'clock, the evening a lovely summer amber, when Simmons docked the ship at the foot of 23rd Street. Menard and Carroll thanked the captain for his great skill in getting them through the storm, then set off in search of family and friends.

The scent of putrefaction was overpowering.

Silence

THE TRAIN LEFT Houston at dawn and for the first few miles made easy progress. The grass on the lowlands had been blown flat, the few visible trees stripped of all leaves, but otherwise Col. William Sterett saw little of note. The sky was a pretty mix of clouds and vivid blue, with that washed quality that so often came after a storm. Big dragonflies patrolled the grass.

Sterett, a writer for the Dallas *News,* had been in the newspaper's office on Saturday when its telegrapher reported losing all contact with Galveston. That in itself was not surprising. Telegraph lines were always being blown down, but the telegraph companies were adept at fixing breaks quickly and routing telegrams through alternate pathways. Even minor storms caused communication to suffer. What made the silence at Galveston so troubling was its duration. The last telegram had come on Saturday afternoon. Now it was Tuesday morning and the lines were still down.

Wild stories had filled the silence. There was talk, clearly exaggerated, that the storm had submerged the entire city under a dozen feet of water at a cost of a thousand lives. Saturday evening someone in Galveston managed to cable a report via Mexico to a resident of San Antonio, notifying him that the storm had drowned his brother. On Sunday a small party of exhausted men from Galveston had arrived in Houston estimating five hundred dead, surely another exaggeration. At a fundamental level, however, all the rumors and reports agreed on one thing:

A powerful storm had struck Galveston without warning and done the city great damage.

Soon Sterett would see for himself. He was riding in a crowded passenger coach attached to a Great Northern relief train bound for Virginia Point, the last railroad stop on the mainland. As one of the region's best-known newsmen, and a Civil War veteran, Sterett had experienced no difficulty gaining permission to board the train, nor had his friend, Tom L. Monagan, dispatched by an insurance company to assess the damage to its interests in Galveston. In Houston, Monagan had volunteered to help prepare the train for departure and was assigned the task of making sure that everyone on the train had an official pass. Relief officials did not want any sightseers sneaking aboard. The train carried soldiers and two commanders: Brig. Gen. Thomas Scurry, adjutant general of the Texas Volunteer Guard, and Gen. Chambers McKibben, commander of the Texas Department of the U.S. Army. It also carried ordinary citizens, and Sterett knew just by the look in their eyes that they had families in Galveston.

At first Sterett and the other passengers joked and talked of minor things, but soon dread filled the car. The wounds in the landscape became more evident. Here and there a house rose from the grass at a cockeyed angle, its curtains blowing free through jaws of fractured glass. The swollen bodies of drowned cattle lay in the pampas like huge black balloons. As more and more debris appeared along the right-of-way, the passengers grew quieter and quieter. In places water covered the tracks. The train slowed until it seemed to make no noise at all. The slowness amplified the dread. For Sterett it brought to mind a funeral cortege.

Masses of lumber appeared along the railbed. Sterett saw fragments of houses, lace curtains, armoires, bedposts, sheets and blankets. He saw boats, and in the distance, a large ship aground on the prairie. A child's rocking horse stood by itself on a low rise, no house in sight. "And so help me," Sterett said, "I would rather have seen all the vessels of the earth stranded high and dry than to have seen this child's toy standing right out on the prairie, masterless."

Debris and flooding forced the engineer to stop the train just north of Texas City, well shy of Virginia Point. The passengers set out on foot. Sterett and Monagan took off their shoes and rolled up their pants, exposing legs so pale as to be nearly translucent.

Now they saw things they had missed from the train. Intimate debris. Stockings, letters, photographs. Their first corpses. What was so striking about the dead was their battered condition. Their bodies had been stripped naked.

At Texas City, the generals seized a lifeboat from the *Kendal Castle,* a British ship blown ten miles from its Galveston pier. They loaded it with soldiers and supplies and began at once to row across the bay, leaving Sterett and the other passengers behind.

Sterett and Monagan spotted a sailboat making slow progress toward the city in a calm that had made the bay "as gentle as a country pond." While waiting, Sterett roamed the bay shore. Where the water met the prairie he saw bloated horses and cows, chickens, cats, dogs, and rats. "Everything, it seemed, that breathed, was there, dead and swollen and making the air nauseous. And by their sides were people."

Groups of men moved along the bay shore hauling bodies from the water and burying them in shallow graves. They buried fifty-eight that day. Sterett found a letter and read the first line, "My Darling Little Wife," then closed it and dropped it back in place.

The sailboat proved to be a large schooner. Monagan, using his authority as an officer of the train, commandeered it and invited one hundred passengers aboard, many of them Galveston residents trying to get back home. It was late Tuesday afternoon by the time the schooner set sail for Galveston. The lack of wind made the journey slow and hot, and all the while the craft moved through a macabre floe of debris. Bodies bumped against the hull. "It must have taken us from four to four-and-a-half hours to get within a half-mile of the city," Monagan said. "It was dark then, pitch dark." They saw only one light on shore.

The generals in the *Kendal Castle*'s lifeboat found the going just as slow, just as bleak. "I am an old soldier," General McKibben said later. "I have seen many battlefields, but let me tell you that since I rode across the bay the other night and helped the man at the boat to steer to keep clear of the floating bodies of dead women and little children, I have not slept one single moment."

As the schooner approached Galveston, the scent of death became overpowering. At one point Sterett looked over the side and saw a dead woman staring back, her face lit by the moon. Some passengers climbed ashore, the rest, including Sterett and Monagan, decided to spend the night aboard the schooner. The captain sailed 150 yards back into the bay and anchored. It was a night, Monagan remembered, "of horrible sounds."

At daybreak, the schooner sailed to the foot of 23rd Street, three blocks due north of Isaac Cline's office. Sterett and Monagan believed themselves to be among the first outsiders to enter the city. They stopped a man hurrying by who told them thousands of people had been killed, so many that disposal crews known as dead gangs had begun burning bodies where they found them.

Sterett refused to believe it. "Surely the man must be mistaken," he told Monagan. "It is always the rule to exaggerate these calamities and he is only repeating what some one has told him."

The two men moved on into the city.

Searching

Isaac stepped outside into a gorgeous dawn, the sky like shattered china. A fast breeze blew the clouds north and brought him the scent of the sea. The morning was cool and bright, bordered to the east by a cantaloupe sky. It was, he said, "a most beautiful day."

In the new light, he saw that the house in which he and his daughters had found shelter was one of the few still standing. A sea of wreckage spread in every direction. Houses had disintegrated. He looked for landmarks and at first saw none, but as his mind adjusted to this new landscape he began to pick out the ruins of familiar structures. The big Bath Avenue Public School, which his children had attended, stood three blocks east, one wing crushed and exposing a large classroom whose floor now hung over the street at a forty-five-degree angle, with thirty-eight desks still anchored in place.

He guessed that the house was located at 28th and P, which put it about three blocks northwest of where his own home had stood, at 25th and Q. When he looked toward his neighborhood, he saw nothing. The pretty Neville house was gone. So was Dr. Young's. His own lot had been scraped clean. And beyond that, where Murdoch's and the Pagoda had stood, he saw only open sky.

Behind him, the bells of the Ursuline convent rang out to summon parishioners to mass. He climbed a mound of debris. The convent, three blocks north, was still standing, but now it looked huge and strange, a feudal castle over a moor of broken wood. The bells were reassuring. With so few houses to absorb the sound, they rang with far greater clarity.

In the wreckage, he saw striped dresses, black suits, black hats, straw boaters. He looked more closely. Some of the clothing covered battered limbs. The dead lay camouflaged under bruises, mud, and shredded cloth, but having spotted one corpse, he now saw many.

Throughout Galveston, men and women stepped from their homes to find corpses at their doorsteps. Bodies lay everywhere. Parents ordered their children to stay inside. One hundred corpses hung from a grove of salt cedars at Heard's Lane. Some had double-puncture wounds left by snakes. Forty-three bodies were lodged in the cross braces of a railroad bridge. "There were so many dead," said Phillip Gordie Tipp, eighteen at the time, "you would sink into the silt onto a body at every other step." He had reached Galveston Sunday morning aboard a small sailboat. "We kept running into so many dead bodies that I had to go forward with a pike and shove the dead out of the way. There was never such a sight. Men, women, children, babies, all floating along with the tide. Hundreds of bodies, going bump-bump, hitting the boat."

Isaac first secured temporary care for his children—perhaps through a friend, or through his church—then made his way to the Levy Building. Blagden was gone. Isaac assessed the damage. Every window had been blown out. Debris was strewn throughout the office. Rain had warped the wood planks of the floor. He climbed to the roof and found it stripped clean of instruments. He surveyed the city. Paving blocks littered the streets. The wharf front was a tangle of masts and rigging, although the big grain elevator seemed little damaged. Steamships once tightly married to the wharf had disappeared. Far down the coast, where Isaac should have been able to see the barest outline of the St. Mary's Orphanage, there was now just a long white arc of beach.

Wagons passed below, headed north. Limbs protruded from under canvas tarpaulins.

Isaac checked the city's hospitals to see if they had survived, and whether anyone inside had seen his wife. The hospitals had weathered the storm well. He may have returned with his injured daughter. At the hospital he heard that a temporary morgue had been established on the north side of the Strand, between 21st and 22nd. He went there next.

The scent of putrefaction and human waste was at once sickening and heartbreaking. It made his loss seem more definite and filled him with sorrow. The warehouse was a large chamber with a ceiling supported at intervals by fifteen-foot iron pillars, between which the dead lay in rows that stretched from wall to wall. Men and women moved intently among the rows as if hunting bargains at a public market. Many bodies were uncovered, others lay under sheets and blankets, which survivors peeled back to expose the faces underneath. J. H. Hawley, an agent for the Great Northern Railroad, saw the faces of many friends. Under one he found the body of a Mrs. Wakelee, "with a faint smile on her lips. . . . her gray hair all matted and streaming in disordered confusion about her shoulders." He saw his friends Walter Fisher and Richard Swain. Lacerations, bruises, and bloating distorted features and made it hard to tell people apart, even whether a man was black or white. The sun warmed the room, accelerating decomposition. "Odors arise," Hawley said, "making it most unbearable."

A photograph survives. It shows at least fifty bodies. In one row, two boys lie side by side. They could be twins. They wear matching shirts. One lies in the fetal position that young children often adopt when they sleep, but his neck appears to be broken. He looks upward over his right shoulder at an impossible angle. His brother watches with a frown. No one wears shoes; no one seems at peace. Many of the dead have the same expression, as if dreaming the same awful thing. Their brows are furrowed, their mouths perfect circles. They could be gasping.

Isaac, moving systematically from body to body, saw men and women he knew or at least recognized—perhaps even some of those who had taken shelter in his house. (Of the fifty, he would learn, only eighteen survived.) He looked for his wife and Bornkessell and the Nevilles, and perhaps Dr. Young. He found none.

There was hope, still, but Isaac was a scientist. Sunday he gave Cora's name to the Galveston *News* as one of the dead. The newspaper came out later that day as a one-page handbill the size of a letter, consisting entirely of a list of people believed dead. Her name was there.

Even Isaac did not yet understand just how lethal the storm had been. For all he knew, the fifty bodies in the morgue represented the majority of those lost. That morning Father James Kirwin, a priest at St. Mary's Church, took a walk through the city trying to come up with an accurate estimate of the dead, then made his way to the wharf, where a group of men were preparing to set off in Col. William Moody's steam yacht, the *Pherabe,* to seek help from the outside world. Kirwin offered the men some advice: "Don't exaggerate; it is better that we underestimate the loss of life than that we put the figures too high, and find it necessary to reduce them hereafter. If I was in your place I don't believe I would estimate the loss of life at more than five hundred."

FAMILIES TALLIED THEIR losses. Anthony Credo learned he had lost nine members of his family. He found Vivian's body near the place where the family's raft had landed, and buried her. But he learned also that his daughter Irene had died along with her new baby and her two-year-old son. His daughter Minnie had disappeared, with her husband and their two boys. So had his eldest son, William, who had spent Saturday at his fiancée's house. Raymond lay badly injured. Soon after Ruby Credo stepped outside on Sunday morning, she saw her first body: Mrs. Goldman. The woman still wore the clothing Ruby's mother had given her when she and her son arrived drenched at the family's house.

Judson Palmer lay in a nun's cell within the Ursuline convent, dressed in a shirt and skirt given him by the sisters. The sisters gave refugees whatever dry clothes they could find. Throughout Saturday night, survivors turned gratefully toward a particularly solicitous—and tall—nun, only to find themselves staring into the stubbled face of a man in a nun's habit.

Palmer drifted in that sad, empty place where hope and grief intertwine. Later a colleague, Wilber M. Lewis, state secretary of the YMCA, wrote to Palmer's friends to tell them the tragic news. "Mrs. Palmer's body was found and recognized the next day. . . . If Lee's body was ever found it was beyond recognition."

As for Palmer: "He was badly bruised by floating debris, but as far as can be seen was not injured internally. His clothes were torn completely off. His mental condition is the most serious now, but we hope for the best."

An eerie peace settled over the city. People bore their losses quietly. John W. Harris was seven when the storm struck, but remembered vividly how the mayor himself paid his father, John junior, a visit on Sunday morning at their house on Tremont Street. One of the finest homes in the city, it had weathered the storm so well that the Harrises had no conception of the devastation elsewhere in town. They were eating breakfast when the mayor arrived. "John," the mayor told the elder Harris, "your whole family is destroyed."

Harris had lost his sisters and their families. Eleven men, women, and children. His son saw him cry for the very first time.

When Clara Barton arrived the next week, she found the silence striking. People moved as if dazed, she said; there was "an unnatural calmness that would astonish those who do not understand it." People grieved, but without demonstration. "You will hear people talk without emotion of the loss of those nearest to them," Father Kirwin said. "We are in that condition that we cannot feel."

Everyone in Galveston experienced some degree of loss; the lucky ones suffered only material damage. Dr. Young had lost his home, but his family had gotten his message and was safe in San Antonio. The Hopkins family too survived, although at first Mrs. Hopkins seemed not to appreciate her good fortune. When the sun came up, she saw that her house, the family's main source of income, had been destroyed. The kitchen, dining room, and two upstairs bedrooms had tumbled into the yard next door. Louise Hopkins would never forget the despair in her mother's voice. "Oh God," her mother said, "why couldn't we all have gone with it."

THE *LOUISIANA* SURVIVED the storm with its cargo badly shifted. Captain Halsey docked briefly in Key West so that his crew could reposition the load, then continued the voyage to New York, where the ship was met by reporters anxious to learn of his encounter with the great hurricane. The storm, Halsey told *The New York Times,* had baffled description.

ISAAC SEARCHED FOR his wife. A photograph exists of what once was his neighborhood. Taken by someone standing near the Ursuline convent and looking south, it provides a view very much like what Isaac must have seen when he emerged from the house at 28th and P on Sunday morning. The ruins of the Bath Avenue Public School stand to the left. Where his house should be, there is only a plain of lumber.

At first glance, the wreckage in the foreground seems to be a homogeneous mass of wood reaching all the way to the horizon, where a pale line demarks the Gulf of Mexico. Close examination with a magnifying glass, however, reveals the base of a wooden swivel chair, a wicker seat-bottom, a steamer trunk, and a surprise. At the right, about where Isaac and his family came to rest, four men stand amid the wreckage. Three are in shirtsleeves and appear to be digging. The fourth stands nearby, watching closely. This man looks like Isaac. Impossible, of course. But he is Isaac's height, has Isaac's small beard. Despite the heat, he wears a suitcoat and hat.

As Isaac searched, he encountered other men and women hunting for their families and friends. They traded information: a woman found here, a man there, a large collection of corpses down near the beach. It is possible that during his search, Isaac encountered a Houston man named Thomas Muat, who came to Isaac's neighborhood looking for his own daughter, Anna, eighteen years old. She had arrived in Galveston a week earlier to visit friends and was staying at the home of David McGill, at 26th and Q, one block west of Isaac's house. McGill was a friend of the Muat family.

The Muats had expected Anna home on Sunday night, but that afternoon learned that no trains had been able to leave Galveston. After a long, anxious night with no word from his daughter, McGill resolved to go to Galveston first thing Monday morning. He and his brother-in-law and two other men boarded one of the first trains that tried to reach Galveston, but got only as far as La Marque, near Texas City. They continued on foot to Virginia Point and there got some disconcerting news: Already that day, the men of Virginia Point had buried two hundred bodies that had drifted across the bay from Galveston.

Muat and his companions used copper wire to lash together three fallen telegraph poles, then hammered a board across the top for a platform. They launched their raft beside one of the railroad causeways and pulled themselves along from piling to piling. Three times the raft capsized. Three times they righted it and moved on, until finally, as daylight faded, they reached the wharf. "What we experienced beggars description," Muat said. "We had to walk over human bodies, cattle, broken box cars and barbed wire, reaching the city about 8 o'clock."

Too exhausted to search, they managed to find a boardinghouse still in operation. Early the next morning, they set out for 26th and Q, and soon found that the McGill house had been "swept out of existence."

They searched further and located McGill's wife at a house a dozen blocks away. The last she saw of Anna, she told Thomas, was after the house had broken apart. Her husband and Anna had wound up on one segment of roof, Mrs. McGill on another. Anna had cried for help, but Mrs. McGill could do nothing. She had not seen them since.

"The only hope we have," Muat said, at the time, "is that my daughter may have been picked up here and is not yet in a condition to tell."

In the absence of a body, there was always hope. Isaac continued his search. But as conditions worsened—as fears of disease grew and as more and more corpses turned up (among them Anna Muat's)—the hunt for miracles and bodies became more complicated. Hope receded, and simple emptiness took its place.

Tuesday, September 11

I. M. Cline, Local Forecast Official, still unable for duty.

"Not Dead"

EVERY DAY, THE editors of the Galveston *News* removed a few people from the list of the dead and placed them on a much shorter list titled "Not Dead." Misreporting had become a problem. The *Tribune* ran a short item under the headline "Be Careful," which urged anyone reporting a death to make certain the victim really had expired. "Several names were turned in as dead and the parties were very much alive."

The list of the dead grew longer and longer. There were so many bodies that disposal became the top priority of the city's Relief Committee, which now governed the city and had appointed subcomittees to manage burial, finance, hospitals, and other tasks. On Monday the burial committee resolved to begin burying the bodies at sea. All day long, fire wagons, hearses, and cargo drays hauled stacks of bodies to the city's wharf, where crews loaded them onto an open barge. The city's racial harmony began to erode. Soldiers rounded up fifty black men at gunpoint and forced them onto the barge, promising whiskey to help make the task of loading, weighting, and dumping the bodies more tolerable.

The day was hot. The barge was moored near the *Pensacola,* whose crewmen stood at the rail and watched intently despite the grotesque images and smells. Workers threw the bodies into the hold with little regard for modesty, until the bodies formed a tangle of swollen buttocks and rigid limbs. One body stood out. It was long and slender and wrapped ever so carefully in white linen. Someone had laid it on a portion of the deck that kept it raised above the other corpses, where it gleamed in the bright sun like a statue of white marble.

By late afternoon, the barge contained seven hundred corpses. A steam tugboat towed the barge to the designated burial ground eighteen miles out in the Gulf, but it arrived well after nightfall and the darkness made it impossible for the crew to work. They spent the night among arms and legs brought back to life by the gentle rocking of the sea. Dead hands clawed for the moon.

At dawn the men began attaching weights to the bodies—anything that would sink. Portions of rail. Sash weights from windows. They worked quickly. Too quickly, apparently, for by the end of the day bodies began returning to Galveston. The sea drove scores of them back onto the city's beaches. Some had weights attached; some did not.

The burial committee found its choices limited. The morgues were already full, burial at sea clearly had not worked, and decomposition was making the bodies hard to handle. The whole business of carting corpses through the streets of the city was itself taking a toll. "It was realized," wrote Clarence Ousley, of the *Tribune,* "that health, even the sanity of people in the streets, forbade the ghostly parade of carts to the wharf, and the only course was to bury or burn on the spot."

The fires began almost at once and for Isaac and thousands of other survivors the quest to find the bodies of loved ones became nearly impossible. The scent of burning hair and flesh, the latter like burnt sugar, suffused the air. Phillip Gordie Tipp's crew, managing a pyre at 25th and Avenue O½, burned five hundred corpses. The city's lifesaving squad, led by Capt. William A. Hutchings, superintendent of the Eighth U.S. Life-Saving District, found and buried 181 bodies, and stumbled across an occupied coffin that had been shipped to the Levy livery company from New Orleans the day of the storm. They buried it too.

The dead gangs worked thirty-minute shifts, and in between were allowed all the whiskey they needed to keep going. "The stench from dead people and animals was so great that they couldn't work longer," one witness said. They worked in long sleeves and jackets and mohair pants, but did not let their discomfort show. They left their noses exposed.

Burning did not seem much of an improvement over the parade of corpse-filled wagons. The idea of burning the bodies of men, women, and children—especially children—was jarring. It seemed like sacrilege. Cremation as a routine mortuarial service was itself a brand-new idea in America. In Galveston the fires were everywhere. Emma Beal was ten at the time of the storm, but watched a dead gang burn bodies at 37th and Avenue P, right near her house. As one body entered the fire, an arm shot up as if pointing into the sky. Emma screamed, but kept watching, and paid for it with nightmares that left her writhing in the dark.

Isaac Cline moved through an increasingly hellish realm. He could not escape the pyres. There was Phillip Gordie Tipp's fire at 25th and O½, another at the foot of Tremont Street opposite the wharf, where several hundred bodies stacked four and five feet deep were burned at once. Fires burned along the beach at intervals of three hundred feet. At night the fires lit the horizon in all directions, as if four suns were about to rise. The men tending the fires soon lost any sense that they were doing something extraordinary. One survivor said the fires became "such a usual spectacle as to create no comment."

Rumor and apocrypha supercharged the night. There was talk that a second huge storm soon would arrive. Isaac's bureau quashed it. On Sunday night, September 16, an immense fire destroyed the Merchants and Planters Cotton Oil Mill in Houston, lighting the sky to the north. In Galveston, where local flames already rimmed the night, it seemed as if the end of the world had come. And for William Marsh Rice, the elderly New York millionaire who owned the plant, it was indeed the end—the hurricane and fire prompted him to begin preparations for transferring a large amount of cash to Houston to begin reconstruction, which in turn caused his valet and an unscrupulous lawyer to accelerate their ongoing plot to poison him.

Black men were said to have begun looting bodies, chewing off fingers to gain access to diamond rings, then stuffing the fingers in their pockets. The nation's press took these stories as truth, then pumped them full of even more lurid details. On Thursday, September 13, the Mobile, Alabama, *Daily Register* told its readers that fifty Negroes had been shot to death in Galveston. "The ghouls," the newspaper reported, "were holding an orgie over the dead."

Nothing of the sort happened, although it is likely that some theft by whites and blacks alike did occur. A reporter for the Galveston *News* reported a rumor that seventy-five looters had been shot in the back, but he was skeptical. "Diligent inquiry discloses the incorrectness of this report." He hedged, however. If any had been killed, he wrote, certainly the total could not have exceeded half a dozen. John Blagden, who survived the storm without injury, heard a rumor that four men had been shot on September 10. "I do not know how true it is," he said, "for all kinds of rumors are afloat and many of them false."

A fog of putrefaction and human ash hung over the city. The steamer *Comal* arrived on Monday and berthed at Pier 26, but her captain was so repulsed by the stench, he moved the ship down the wharf. The weather was clear and bright, and hot. "Fearful hot," one man said. He estimated the temperatures at close to 100.

"Every day the stench from rotting bodies got worse," Ruby Credo said. "I could barely keep from retching, it was so bad."

ISAAC READ THE *News* closely. Everyone did. In the days after the storm it became the city's main source of information about friends and relatives. On Thursday, it looked like a real newspaper again. It was back to full size and gave readers its first big narrative of the storm, along with a drizzle of little stories, including a report that someone's pet prairie dog had been rescued alive from a dresser drawer. The paper also ran a list of telegrams that had accumulated at the Western Union office on the Strand, undelivered because so many recipients and messenger boys were dead, and because 3,600 homes had disappeared. The list of telegrams had five hundred names.

On Friday the *News* ran its first advertising since the storm. Companies reassured customers they would not jack up prices to take advantage of the disaster. A store called The Peoples offered goods at manufacturers' cost. Friday's death list wrapped around an ad for the Collier Packet Company, which offered coffeepots, cups, clotheslines, brooms, rakes, shovels, nails, lanterns, lamps, and soap, and promised, "Positively no advance in prices." An ad for H. Mosle and Company offered Tidal Wave Flour at one dollar a sack.

The death list took up a full page and a fraction of the next, and included fragments of information that telegraphed to readers larger truths about the disaster. Black victims were identified as colored. The list provided vivid evidence that the storm had crossed all lines of race, profession, and class. It killed steamship agents, mailmen, longshoremen, a prizefighter, a deputy U.S. marshal, and thirteen unidentified inmates of the Home for the Homeless. It killed twenty-two people at the residence of "Francois, a well-known waiter," and pruned to a stalk the family tree of the Rattiseau clan, killing Mrs. J. C. Rattiseau and her three children; J. B. Rattiseau, his wife, and four children; and C. A. Rattiseau, his wife, and seven children. It drowned Mr. and Mrs. A. Popular and the four Popular children, Agnes, Marnie, Clarence, and Tony. It killed Sanders Costly and Clara Sudden, Herman Tix and H. J. Tickle. It killed John Grief and the entire Grief family.

The list included a man named Pilford of the Mexican Cable Company and his four children. The place of death, the entry said, was "Twenty-fifth and Q." Isaac's corner. Perhaps even his house.

On Friday the newspaper returned to its practice of running "Personals," but now these took on a rather different character.

W. M. R. Clay placed a notice to the attention of Jetta Clay. "I am here," it said, "2002 L. Come at once."

Charles Kennedy placed one seeking Fred Heidenreich. "If alive, come to 24th and Church. Your brother Ben is there."

The following Tuesday, a query read: "RYALS—If Myrtle, Wesley, Harry or Mabel are living, please address their mother, Mrs. Ryals, 2024 N."

HELP BEGAN ARRIVING. The Army sent soldiers, tents, and food. The train-ferry *Charlotte Allen* brought a thousand loaves of bread from Houston. The steamer *Lawrence* brought one hundred thousand gallons of fresh water. The Grand Dictator of the Knights of Honor arrived to look after the needs of his Galveston brethren. Clara Barton arrived to look after everyone, and immediately telegraphed home: "Situation not exaggerated." She had expected many orphans, but found few. The storm had been hardest on the small. She came with a trainload of carbolic acid and other disinfectants supplied by Joseph Pulitzer's New York *World*. William Randolph Hearst's New York *Journal* sent a train too. It left first but arrived last. Which peeved Hearst no end. He dispatched one of his top writers, Winifred Black, his famed "sob sister," whom he had brought to New York from San Francisco specifically to battle Pulitzer. The storm, she found, had unearthed a Galveston cemetery. The *Journal*'s headline shrieked: "EVEN THE GRAVES GIVE UP THEIR DEAD."

The great hurricane dominated the front pages of newspapers from Miami to Liverpool and generated a tidal wave of contributions, most channeled through the Red Cross. Hearst, in the name of an outfit called the New York Bazaar for Galveston Orphans, gave $50,000, a fortune. In his role as publisher of the *Journal,* he gave $3,676.02. The Kansas State Insane Asylum sent $12.25. The Colored Eureka Brass

Band of Thibodaux, Louisiana, sent $24. The Elgin Milkine Company of Elgin, Illinois, sent seventy-two bottles of its dried-beef tablets and powder. The tablets came in lemon and chocolate. The Fraternal Mystic Circle, Elmwood Ruling, No. 430, of Gainesville, Texas, sent $50. The Ladies of the Maccabees of the World, Sacramento Hive No. 9, sent $329.25. The city of Liverpool gave $13,580, the Cotton Association of Liverpool, $14,550. In the United States, the state of New York sent the most money, $93,695.77. New Hampshire sent the least—a buck—matching the contribution of Moose Jaw, Canada. The Sabbath School of Odell, Illinois, sent $4.10 for the few orphans Barton did find, and got a warm personal letter in return. "It would not surprise me if in its careful expenditure there were not a few playthings," Barton wrote, "possibly a doll, a wooly dog, an antelope or a little village."

Among the contributions that moved her most was $61 from workers at the Cambria Steel Company, Johnstown, Pennsylvania. They made no mention of the ordeal they had gone through eleven years earlier after the failure of a dam at a rich-man's club high above town.

Observers within the Weather Bureau contributed to a fund for the relief of their Galveston colleagues, earmarking $200.76 for Isaac, $150 for Joseph, and $50 for John Blagden. Isaac sent his deepest thanks. A rather unctuous letter went to Willis Moore from an observer in the West Indies Service, William H. Alexander. Alexander did not contribute to the Galveston relief fund, but professed to feel deeply for the station and for the state of Texas. "So, feeling thus and fearing lest my silence be attributed to indifference, I felt that in justice to myself I should state that I sent to Galveston to a needy friend as soon after the storm as possible the sum of $11.00 which was every cent that I felt able to contribute."

There is no record of any contribution from his Indies superiors, William Stockman and Col. H. H. C. Dunwoody.

• • •

ISAAC RETURNED TO work on Monday, September 17. What he had done during his eight days away from the office is unclear. One local historian believed he was in the hospital recovering from his injuries, but this seems unlikely. Isaac was not seriously injured, at least not physically. He continued to file telegrams to Washington. Most likely he spent this time struggling with the hunt for his wife, the care of his children, and his own grief. There was much for him to do. He needed to find a permanent home for his children and a woman to care for them. He put Joseph in charge of the office, although it must have pained him to do so. Joseph reveled in his new command, and in his brother's absence. Telegrams from the bureau became more dramatic. At 11:30 A.M., Tuesday, September 11, Joseph fired off a breathless telegram to Moore, in which he reported that Bornkessell was still missing, Isaac had been injured but "not seriously," and "nearly half the city" had been washed away. "I am badly injured. Two thousand dead found burying at sea."

Exactly three minutes later, a more businesslike telegram entered the wires, this composed by Isaac. "All mail communication cut off since noon Saturday. Can get no material on which to base crop reports. All messages sent by boat to Houston. Instruments erected temporarily by Blagden. J. L. Cline still on duty but unable to do much."

These were hard days for Isaac. He believed in work and in filling his day to the limit with productive effort, but in so doing, he had put love and family in a box that he had allowed himself to open only rarely. A mistake, he saw now. He had lost his wife and nearly lost a daughter. How completely Cora had held his world together now became apparent to him. His children needed food, warmth, a dry place, and most of all they needed him. As the city had fallen, so had the neat compartments of his life.

His house had disappeared, along with everything that described his past—all his photographs, letters, his beloved Bible, and the manuscript of his nearly finished book on climate and health, the second time the book had been destroyed. His station was in disarray. Kuhnel had deserted. Baldwin, on mandatory furlough, had gotten safely away just a week before the storm.

And Bornkessell was surely dead. He had just built a home in the city's West End, but searchers found only empty ground. Neighbors apparently had sought shelter in his house, placing in him the same faith others had placed in Isaac. On the morning Isaac returned to work, he read in the Galveston *News* a query from a Houston man named Harry M. Perry. "I wish to report to you as among the missing and undoubtedly lost my wife and son Clayton, aged 7. They were visiting Mr. and Mrs. Theodore C. Bornkessell, who resided in their new cottage on the north side of the shell road, about a mile west of the Denver Resurvey. I reached Galveston on the first trip of the steamer *Lawrence* and searched the ground carefully from the site of the house to the bay, but could find no trace of them. Everything out there went straight into the bay, as there was nothing to stop it. The house is entirely gone, but some of its wreckage is lodged in trees a mile northwest. My wife was about 5 feet 5 inches tall, wavy, medium length black hair, 30 years old, looking younger, but hair had many gray ones in it. . . . Should any record of such persons have been made by any one it is needless to say I will appreciate all possible information. Mr. and Mrs. Bornkessell were undoubtedly lost with them."

As far as the station was concerned, things could have been worse. Joseph was indeed injured, but not as badly as he seemed to think. He had never dealt well with injury or illness. Blagden, luckily, was well and full of energy. The Levy Building was still sound.

Isaac could not help it, but now and then a thought whispered through his mind that he should have come to the Levy Building with his family, instead of trying to weather the storm at home. Why had he chosen that course? Was it pride? For the sake of appearances?

Joseph, underneath his demonstrations of sympathy, seemed all too aware that he had called it right and Isaac had not.

There were dreams. Isaac fell asleep easily each night and dreamed of happy times, only to wake to gloom and grief. He dreamed that he had saved her. He dreamed of the lost baby. "A dream," Freud wrote, in 1900, in his *Interpretation of Dreams*, "is the fulfillment of a wish."

During the week he worked on his official report on the storm. Psychically, it was a difficult task. His wife was still missing. The air stank of rotting flesh and burned hair. Always in the past he had been able to separate himself from the meteorological events he described. Hot winds. Paralyzed fish. He was the observer looking upon these phenomena through glass. But this storm had dragged him to its heart and changed his life forever. As he sat down opposite his typewriter, human ash dusted each fresh sheet of paper.

He began: "The hurricane which visited Galveston Island on Saturday, September 8, 1900, was no doubt one of the most important meteorological events in the world's history."

There was so much he wanted to say, but could not—how headquarters and the West Indies Service had failed to recognize the storm as a hurricane, how even he had not understood the signs of warning until too late. That was the most difficult part. He could not describe these conjoined failures, for to do so would have been to damage the bureau in its struggle for credibility.

Instead, he wrote: "Storm warnings were timely and received a wide distribution not only in Galveston but throughout the coast region."

He left out the specific character of these warnings, and the fact that none mentioned a hurricane.

As he wrote, a question consumed him: Why did so many people die? No previous storm on the U.S. mainland had come even close to causing such loss. Why this one? Was he, Isaac, partly to blame?

As if to address the question, he described how on Saturday morning he began warning the public to seek a safe place to spend the night. "As a result thousands of people who lived near the beach or in small houses moved their families into the center of the city and were thus saved." In later years, the number of people he claimed to have warned increased to twelve thousand.

Isaac struggled also with how to tell the story in a dispassionate, scientific way, and bleach it of his personal experience. He found this impossible. This was *his* storm. What he knew of it came from living through it.

In a few austere paragraphs, he described the collapse of his house and his night on the wreckage. He devoted a single line to Cora's apparent death: "Among the lost was my wife, who never rose above the water after the wreck of the building."

His account was spare, nothing like the florid writing so common in his time. For him, however, it was new. He had never written an official document in the first person before, only in the passive voice of a bureaucrat; certainly he had never mentioned his family. It was risky. He was violating an unwritten tenet of bureau culture as it had evolved under Willis Moore: Do not ever let your own star shine more brightly than the chief's.

But there was no other way to tell the story. Isaac sent it to Moore with a cover letter in which he wrote, "My personal experience was so interwoven with the progress of the storm that it appears that I should include it in the report. If it should not be embodied in the report please omit that portion.

"Very respectfully,

"Your obediant servant,

"I. M. Cline."

A Letter from Moore

ON FRIDAY, SEPTEMBER 28, as hundreds of fires still burned in the city, Isaac Cline read the Houston *Post* and there saw an angry letter from Willis Moore defending the Weather Bureau's performance in forecasting the hurricane. The letter troubled Isaac. Moore's account veered from reality; why was he changing the story?

Moore had written to the *Post* in response to an editorial that accused the bureau of having failed to predict and track the storm. The editorial stated: "The practical inutility of the national weather bureau, for certain sections of the country, at least, was never so conspicuously shown as on Friday and Saturday last when South Texas was left without any warning of the coming storm, or at least its severity." The editorial then quoted the forecasts for Texas that had been wired from the bureau's Central Office just before the storm. "With the weather bureau saying that Saturday would be 'fair; fresh, possibly brisk, northerly winds on the coast' of East Texas, who would have looked for the most destructive hurricane of modern times on that coast on that date?"

Moore, in a five-page letter, protested that on Friday morning storm signals were raised in Galveston and "a few hours later" were changed to hurricane signals. He called Isaac "one of the heroic spirits of that awful hour," and offered a dramatic, but incorrect, account of Isaac's day. "When the last means of communication with the outer world had failed he, instead of going to the relief of his own family, braved the furies of the storm and the surging waters and, reaching a certain telephone station at the end of a bridge, succeeded in sending out from the doomed city the last message that was received until after the passage of the storm. . . . After performing this service for the benefit of the whole people he returned to his own home, to find it destroyed and his wife and one child lost."

Isaac, at this point, still considered Moore a personal friend. It hurt him, no doubt, that Moore had distorted the story of his experience in the storm. Isaac had lost his wife and home, and had nearly lost a daughter, but Moore could not be bothered with the actual details. What troubled Isaac most was Moore's statement that an order to raise hurricane warnings had been sent to Galveston and that hurricane flags had been raised as early as Friday. It simply wasn't true.

Isaac clipped Moore's letter and an accompanying blurb in which the *Post*'s editors stated they had gladly printed Moore's response because they had no desire "to captiously find fault" with the bureau, adding archly, "We would all rather believe that the weather service was valuable than that it was of no use to the public." Isaac mailed the clippings to Moore that day, with a cover letter that was defensive but also obliquely critical of Moore. Isaac insisted he had done all he could on Saturday. In fact, he told Moore, he had just spoken with an editor of the Galveston *News* who had assured him that his station's warnings had saved "more than 5,000 people."

Isaac ignored Moore's distortion of his personal ordeal, but quietly disputed his claim that the Central Office had ordered hurricane signals raised. "Regarding the warnings received at Galveston I desire to say that the hurricane warning never reached us," Isaac wrote. The last storm advisory received from Washington, he said, was an order that arrived at 10:30 A.M. Saturday specifying only a change in wind direction of an existing storm warning. Always the good soldier, Isaac gave

Moore an escape. "I presume," he wrote, "the hurricane warning which followed a few hours later did not reach Houston until after the wires had gone down."

There was more he wanted to say, but did not. Years later, in a personal memoir, he wrote that the only warning given the people of Galveston came from him and his station, in defiance of Moore's "strict orders" against unauthorized storm warnings. "If I had taken the time on the morning of the 8th to ask for approval from the forecaster in Washington and waited for his reply the people could not have been warned of the disaster." Loss of life, he wrote, "would have been twice as great."

But he conceded he too had underestimated the storm. "I did not foresee the magnitude of the damage it would do."

MOORE CONTINUED TO portray the bureau as having expertly forecast and tracked the hurricane, and credited in particular the West Indies Service. In an article in the October issue of *Collier's Weekly,* one of the most influential magazines of the day, he wrote, "It is a remarkable testimonial to the foresight of the present Secretary of Agriculture, Honorable James Wilson, that the meteorological service inaugurated by him during the Spanish-American war as a protection to the American fleet was, by the last Congress, permanently adopted as a part of our National Weather Bureau, on account of its beneficent application to the peaceful ways of trade and commerce. Without the reporting stations of the new service the Weather Bureau would have been unable to detect the inception of the Galveston hurricane when it was only a harmless storm, and, when it reached the intensity of a hurricane, to issue timely warnings in advance of its coming." He repeated his distorted account of Isaac's ordeal.

Most U.S. newspapers, unaware of the nuances of the bureau's performance and inclined in those days to be more accepting of official dogma, adopted Moore's view. The Boston *Herald* applauded the bureau for its "excellent service." The Buffalo, New York, *Courier* said the bureau's forecasts testified to its "advanced efficiency." The *Inter-Ocean* of Chicago, Illinois, wrote, "Simple justice demands public recognition of the efficiency of the Chief of the Meteorological Bureau and his staff."

Few asked the obvious question: If the bureau had done such a great job, why did so many people die? More people perished in Galveston than in any previous U.S. natural disaster—at least three times as many as in the Johnstown Flood.

SOON AFTER THE storm, Father Gangoite of the Belen Observatory discovered William Stockman's patronizing remarks about how the poor ignorant natives of the islands had become accustomed to learning of storms "only when they were nearly past." Gangoite brought them to the attention of the Cuban press. In the wake of the Galveston storm, Gangoite and Cuba's editors saw the remarks as highly ironic. The *Diario de la Marina* noted that the Cuban public always gave "greater credence" to Gangoite's forecasts, and that the facts justified this attitude.

"An example?" the editors asked. "Here is a recent one. The same day that the Weather Bureau published in the newspapers of Havana that the last hurricane had reached the Atlantic, the Belen Observatory said in the same papers that the center had crossed the eastern portion of the island and that it would undoubtedly reach Texas. A few hours later the first telegraphic announcement of the ravages of the cyclone in Galveston was received."

The editorial concluded: "As this occurrence is very recent it affords a most delightful opportunity for the verification of what has just been published in the U.S., that until the establishment of the Weather Bureau in Havana, forecasts relating to hurricanes were unknown by the people of Cuba."

Six days after the storm, the War Department, apparently fed up with Stockman and Colonel Dunwoody, revoked the ban on Cuban weather cables. Moore was furious. In a letter to the secretary of agriculture, he fumed, "I know that there have been many secret influences at work to embarrass the Weather Bureau. I regret that the restriction that heretofore has been placed on the transmission of private observations and forecasts over the Government lines has been removed." He turned petulant. "It is apparent to me and to every ranking officer . . . in the West Indies that the people do not appreciate our service, that the only thing they want is to kick us and say good-bye."

By way of retribution, he asked permission to halt the bureau's climate and crop service in Cuba and to move the headquarters of the West Indies hurricane network out of Havana. He also wanted authorization "not to issue hurricane warnings to any part of Cuba so long as the War Department permits the transmission over Government lines of irresponsible weather information."

28TH AND P

The Ring

THERE WAS A point where families knew their missing members were gone for good, although different people reached that point at different times. Children reached it last of all. There were miracles still, like Anna Delz, sixteen years old, who had been washed to the mainland and mourned for dead until a week later she finally made her way back to Galveston. Stories like this were heartening, especially if you concentrated on the joy the newfound survivors brought to their families and friends, but they also were distressing, especially for parents who knew their spouses were dead but whose children saw each new miracle as a sign that their own mothers or fathers might also return.

Isaac knew Cora was dead. He knew it on a rational, scientific level. Even so, he needed to find her, lest a part of him always wonder where she was, and a very tiny part of him always wonder whether she was even dead. He needed to find her also for the sake of his children. They still believed their mother one day would walk through the door and scoop them into her arms. Little Esther was the most open about it, wondering aloud when her mama would come home. The eldest, Allie May, tried to act adult and maternal, but Isaac knew that on some level she too believed her mother would come back. The children prayed for this. At night he often woke to hear one or another of his daughters crying in her sleep. Sometimes they cried upon wakening. Freud said, "The dreams of young children are pure wish-fulfillments and are for that reason quite uninteresting compared with the dreams of adults. They raise no problems for solution."

At the office, things quickly returned to normal. Isaac, Joseph, and John Blagden received commendations; Ernest Kuhnel, the deserter, was dropped from the rolls. New instruments arrived and the men returned to making their routine daily observations. Pyres burned everywhere. Work crews erected scores of new homes. The Rollfings found one and moved in. The scent of fresh-cut lumber diluted the scent of death. Cotton began flowing through the port, no doubt to Houston's dismay. Squads of men hacked away at the immense spine of debris that had come to a halt on Avenue Q. What was so striking was the quiet. The men did not have jackhammers and chain saws, of course. Only axes, hammers, handsaws, and crowbars. They burned the wreck in segments, after salvaging intact sinks, lamps, stoves, coffeepots, pans, even commodes, figuring someone might need them. The Red Cross gave out food and clothing, but found much of its supply of donated clothing unusable, either too warm for the climate or too shabby, clearly the discards of distant souls who believed survivors were in no position to be picky. Someone donated a case of fancy women's shoes, but all 144 shoes were for the left foot, samples once carried by a shoe-company traveler. Contributions slowed. Discord rose. Barton was accused of withholding clothing from Galveston's destitute blacks, and of squandering money in payments to members of the Relief Committee. The *Palmetto Post* of Port Royal, South Carolina, called her a vulture. None of it fazed her. The same thing occurred at every disaster she attended. "It is," she wrote, "an unfortunate trait in the human character to assail or asperse others engaged in the performance of humanitarian acts."

Throughout September, bodies emerged from the wreckage at a rate of over one hundred per day. Two hundred seventy-three bodies came forth on September 19. The next day's *News* speculated, "It is possible, but highly improbable, that the list of storm victims will aggregate 6000 souls." As the days passed, identification became impossible, unless the dead happened to wear some clearly distinctive piece of jewelry or clothing.

Toward evening on September 30 a demolition gang assigned to help dismantle the spine of wreckage that still stretched from one end of the city to the other began working in the vicinity of 28th and Avenue P. They took on only a small portion at a time. To think in terms of the whole was simply too disheartening. They saw the world not in acres, but in cubic yards.

As they dug through the rubble the now-familiar scent of decomposition became stronger. None was surprised by this. The spine had proven from the start to be a rich seam of corpses.

The wall of a house had come to rest here. They disassembled it and stacked the reusable lumber and siding in a great pile. Underneath they found a dress tangled in the debris, and within the clothing, the remains of a woman. The woman wore a wedding ring, and a diamond engagement ring.

What happened next is unclear. It is possible something in the debris signaled to the men that the house had belonged to Dr. I. M. Cline, the weatherman, and that the men then dispatched someone to bring him to the scene. It is also possible Isaac was already there, waiting, having long ago considered the possibility that his wife's body might have come to rest near where he and the children had floated to safety. By then Isaac would have established a routine that he followed every day, a scientific approach to the search that began with the *News,* and ended each evening with a tour of likely places where his wife might have lain. He probably justified it as good exercise.

Isaac recognized the ring. Something closed in his heart and a kind of peace rose within him, like a flush of embarrassment. "Even in death," he wrote, years later, "she had traveled with us and near us through the storm."

The work crew did not burn her body—further evidence that Isaac was present during or soon after its discovery. The body was transported to the city's Lakeview Cemetery. On October 4, 1900, as the weather began to cool, Isaac and his daughters, and Joseph, gathered on the cemetery's grounds, at Block 47, Lot E, ½ of 3, and watched as a coffin bearing Cora May Bellew Cline was lowered slowly into the earth.

Isaac kept the ring, had it enlarged, and wore it himself. It was this ring that gleamed like a beacon from his photographic portrait. He wore it also on December 31, 1900, when Galveston prepared to enter the twentieth century. The city looked new. Its streets were clear, the pyres gone. The civilized smoke of steamships now drifted over the city. And the glad scents were back, of coffee and fresh wood, and horses. Music rang from the restored Garten Verein, and from the banquet hall of the Tremont Hotel, and the dance parlor of the Artillery Club. Sad men made love that day in the house across the alley. Beer flowed everywhere, and there was laughter. Children ran along the beach as their parents followed, anxious as always about the sea. And then the rockets came, arcing over the black water of the Gulf in bursts of yellow, red, and gold. Isaac was there with his baby and Allie and Rosemary. Joseph was gone, in Puerto Rico. There was Judson Palmer, alone but among friends. There was Louisa, with August senior and junior and Helen and little Lanta. There was Mrs. Hopkins and her children, and Anthony Credo and his children, Raymond alive, Pearl's arm nicely healed. Voices came next, Isaac's tenor merging with August's and a thousand other voices over the soft whisper of the sea, filling cups of kindness for old times past. And then the ghosts came. They gathered silently on the beach. Cora Cline. Vivian Credo and her sisters Irene and Minnie and her brother William. Little Lee Palmer and his mother, and of course, Youno. Lost families remembered. Tix, Popular, Grief.

That night, New Year's Eve 1900, a piece of very strange news flashed over the submarine cables from England. A wind had risen so freakishly strong it had toppled one of the great pillars of Stonehenge that no wind had budged for ten thousand years. The twentieth century had begun.

PART VI

Haunted

_____

ISAAC

Haunted

The Storm

ON MONDAY, SEPTEMBER 10, Willis Moore telegraphed the New York *Evening World* with a report on the hurricane's travels after it left Galveston. The cyclone, he wrote, "had lost its distinctive character as a destructive storm, and its future energy will more likely be expended in general rains over the western country rather than in high winds."

Once again Moore had let the expected obscure the real. Somewhere in the heavens over Oklahoma, the storm's lingering vortex entered the great low-pressure system then moving eastward across the country. It rapidly regained power and roared north, much to the dismay of A. I. Root, president of a Medina, Ohio, company that sold beekeeping supplies. As early as Monday he watched his personal barometer begin to drop "in a very unusual way," yet all he saw from the Weather Bureau were telegrams forecasting fair skies for Monday and Tuesday, partly cloudy conditions on Wednesday. Instead he got a destructive windstorm that tore his company apart. He wrote to Moore, "Now wasn't it a mistake that there wasn't anything said about the big blow?"

The Central Office countered that it was not bureau policy to send wind forecasts to inland locations.

The storm brought hurricane-force winds to Chicago and Buffalo, this even after crossing America's vast midriff. It killed six loggers trying to make their way across the Eau Claire River and nearly sank a Lake

Michigan steamship. It downed so many telegraph lines that communication throughout the Midwest and the northern tier of the nation came to a halt. On Wednesday night, the storm savaged Prince Edward Island, then burst into the North Atlantic. Manhattan, half a continent south, received winds of sixty-five miles per hour.

As thousands of men moved into the countryside to replant telegraph poles and string fallen cable, reports began to emerge of shipwrecks in the Atlantic. The storm sank six vessels off Saint-Pierre, six more in Placentia Bay, four at Renews Harbor, and drove forty-two fishing boats aground in the Strait of Belle Isle between Newfoundland and mainland Canada. The storm raced in a cold and lethal arc across the top of the world until it fell at last into Siberia and disappeared from human observation.

America cooled. The Cascades grayed under frost. Snow fell on the Wasatch Front east of Salt Lake City. At Sherman, Wyoming, snow accumulated to a depth of thirteen inches. From Chattanooga to Brooklyn, men and women greeted the day with a feeling not unlike love.

Galveston

GALVESTON COUNTED ITS dead. The city conducted a census and in October reported a tally of 3,406 confirmed deaths. Eight of the city's twelve wards had lost 10 percent or more of their residents. The storm killed 21 percent of the Twelfth Ward, 19 percent of the Tenth. The Galveston *News* published its final death roster on October 7, and listed 4,263 names. Early in 1901, the Morrison and Fourmy Company, which published the city directory, conducted its own canvass and found an overall loss in population of 8,124. Two thousand of these had simply moved from the city, the company believed. That left 6,000 dead. Informal estimates placed the toll at 8,000, even 10,000, not including the several thousand deaths that occurred in low-lying towns on the mainland. No one knew how many bodies still rested in the sea. "Many people," one survivor noted, "would not eat fish, shrimp, or crabs for several years."

The city fathers vowed to rebuild. They created an elaborate exhibit for the World's Fair of 1904 to tell the world of the city's great plans to build a seawall and behind it a shining new Galveston. The Galveston Flood concession quickly became one of the most popular exhibits at the fair. An artificial wave machine threw a tidal wave across a tableau of Galveston. The sun rose upon a ruined city. Night fell. The new day saw the ruin replaced by a great gleaming metropolis protected from the sea by a giant wall.

This time Galveston built the wall. It rose seventeen feet above the beach, and stood behind an advance barrier of granite boulders twenty-seven feet in width. *McClure's Magazine* called it "one of the greatest engineering works of modern times." But the city's engineers, among them Colonel Robert, knew a seawall alone was not enough. They raised the altitude of the entire city. In a monumental effort, legions of workmen using manual screw jacks lifted two thousand buildings, even a cathedral, then filled the resulting canyon with eleven million pounds of fill. The task, completed in 1910, had an unintended benefit: It ensured that all corpses still buried within the city remained well interred.

There were moments of brightness. The city built a grand new opera house to replace the one destroyed in the storm. Al Jolson came. So did Sarah Bernhardt and Anna Pavlova. To signal the city's faith in itself, several of its leading citizens built an immense new hotel, the Galvez, right inside the seawall, as if taunting the Gulf with the city's new resolve. Galveston's Relief Committee evolved into a new form of city government, in which the mayor became, in effect, chairman of a board of elected commissioners who each managed a different city function. Reformers saw it as a way of defeating Tammany-style politics, which tended to concentrate power in the hands of a single boss. Hundreds of cities across the country adopted the form. It caused Harvard's president, Charles Eliot, to proclaim the dawn of "a brighter day" for America. "We have got down very low in regard to our municipal governments, and we have got dark days here now, but we can see a light breaking, and one of the lights broke in Galveston."

But the great hurricane—call it Isaac's Storm—had struck with abysmal bad timing. Just four months later, an event occurred nearby that changed the history of the nation, arguably the world. The ranchers of Beaumont, Texas, had long heard how gas and greasy water sometimes bubbled to the surface of a strange knoll in the prairie outside town. A few men hunted oil there and gave up, but others followed, drawn by the stories. On January 10, 1901, a crew working for an Italian immigrant named Antonio Francisco Lucich, self-named Tony Lucas, ran for their lives as thunder roared from their drill tower. Oil had already made a few fortunes in America, but this was different. The place was Spindletop. Lucas had punctured a vast underground basin of oil. The rig spouted America's new gold—but showered the wealth on Houston, not Galveston.

As Galveston grieved and struggled to regain the world's confidence, Houston dredged Buffalo Bayou. Houston was inland, therefore safer, and it was closer to the big cross-country rail corridors. Oil eclipsed cotton. Great ships in black, red, and white glided quietly past Galveston, bound for the wharves of Buffalo Bayou.

A silence settled over Galveston. Its population stopped growing. It acquired all the sorrows of modern urban life, but none of the density and vibrance. It became a beach town for Houston.

Joseph

SOON AFTER THE storm, Willis Moore promoted Joseph to section director, with an increase in salary to $1,500 a year from $1,200, and ordered
him to Puerto Rico to take over the island's weather station. Joseph
dreaded the assignment, claiming his health was not good enough for a
tropical climate. On November 3, 1900, two days before Joseph was
scheduled to leave Galveston, Isaac notified Moore that Joseph "is
unable to leave his room. He has been under medical treatment since
the hurricane and has at last been compelled to take to his bed."

A month later, Joseph, in a letter that dripped reluctance, wrote to
Moore that he was now ready to go. "I believe that I have fully recovered
from injuries of glands and blood vessels in [my] left leg, and as it is the
wishes of the Bureau that I proceed on to Porto Rico, I will do so with
pleasure." He asked Moore, however, to reconsider his transfer if "the
climate there proves adverse."

Joseph did go to Puerto Rico, and in the August 1901 issue of the
Monthly Weather Review, wrote, "The climate is not so oppressive as
one might expect in the Tropics. A cool, very pleasant, and most welcome breeze generally blows across the island, particularly in the afternoon and at night, which adds much to the comfort of the inhabitants."

By then, however, he was already back in the United States. He
had been back for months. In the spring of 1901, Moore at last had
acknowledged Joseph's concerns and on April 5 wrote to the secretary
of agriculture recommending that Joseph be returned to the United
States on account of his "feeble" health. Moore demoted Joseph to his
old rank of observer and cut his salary by $200 a year.

Two weeks later, as if deliberately trying to intensify the rivalry
between Joseph and Isaac, Willis Moore promoted Isaac and ordered
him to New Orleans to take charge of a newly created Gulf Forecast

District encompassing Texas, Oklahoma, Louisiana, Arkansas, Mississippi, Alabama, and the Florida panhandle. He raised Isaac's salary $200 a year, to $2,000.

Moore

IN 1909 IN a widely published forecast Willis Moore announced that the weather for William Howard Taft's inauguration would be "clear and colder."

Snow fell.

Isaac

ISAAC CAME TO see his transfer to New Orleans as punishment for his having become too successful at forecasting frosts, floods, and storms. He believed Moore considered him a threat to his own job. "When a station official performed work that attracted the attention of the public and was commended by the press," Isaac wrote, "Moore frequently sent him to some part of the world where he could not render conspicuous service."

To Isaac, New Orleans was just such a place. It was, he wrote, "a dumping ground for observers who were guilty of drunkenness and neglect of duty and whom it was necessary to discipline." The low level of talent not only made it difficult for Isaac to improve the station's performance, it also forced him to invoke harsh disciplinary measures, which in turn poisoned his own reputation within the bureau. Many years earlier Gen. Adolphus Greely had sent troubled employees to Galveston, but Isaac saw those transfers as good-faith efforts to save careers. Moore, he believed, had other motives. "The object," Isasac wrote, "was to give my station a bad record in dealing with personal problems."

Isaac's disillusionment deepened when Moore pressured him to assist Moore's campaign to become secretary of agriculture under Woodrow Wilson. Moore used bureau officials and bureau time to promote his ambitions and became so convinced Wilson would choose him that he designated a man to take his place as chief.

Wilson picked someone else. The Justice Department launched an investigation of Moore's politicking, and Moore spread the word to Isaac and other officials to destroy all correspondence related to his campaign. At nine o'clock in the morning on April Fool's Day, 1913, an agent with the Justice Department walked into Isaac's office and demanded to see all correspondence between him and Moore. The agent clearly expected Isaac to claim no such material existed.

Isaac believed in loyalty and hard work and in the essential goodness of men, but he had learned much in those thirty years since his first arrival in Washington. He handed the agent a thick file containing all of Moore's campaign directives, complete with postmarked envelopes.

Moore was fired.

The rivalry between Isaac and Joseph evolved into complete estrangement. The clearest evidence appears in a forlorn document deep in the records of the National Archives. It is an account of the Galveston storm that Joseph wrote in March 1922, in which he goes to great, almost comical, lengths to avoid using Isaac's name or even to acknowledge him as his brother. When Joseph describes his own journey to Isaac's house on the Saturday of the storm, he never identifies its owner. It is only "a house" in which fifty people happened to have congregated. "At eight o'clock," Joseph writes, "the house we were in went to pieces, and as the house went over I broke through the window and climbed on the side of the framed house, and carried two children to safety. . . . Finally the house went to pieces and a short distance away I observed 3 others coming out of the water. These 3 were also saved."

On the night of the storm the lives of Joseph and Isaac touched with an intensity that only a man blinded by anger could disavow. Perhaps Joseph resented Isaac's subsequent success within the bureau, or Isaac's failure to contest Moore's portrayal of Isaac as the great hero of the storm. And maybe Isaac, for his part, transformed his own guilt into a perverse anger at Joseph for having been right about urging everyone to evacuate. Maybe each time Isaac saw Joseph the magnitude of his own error came roaring back to him.

Maybe Joseph sensed this, and played to it.

The hurricane changed Isaac. He gave up the study of climate and health and concentrated instead on trying to find out why the storm had been so deadly. He wrote two books on hurricanes, thus fulfilling his childhood dream of writing an important scientific treatise. He became one of the nation's leading hurricane experts. It was Isaac who established that a hurricane's deadliest weapon was not direct wind damage, as bureau dogma held, but its wind-driven tide, and that this tide provided important warning signals. He was not shy about taking credit. In his monograph "A Century of Progress in the Study of Cyclones," published in 1942, he wrote, "I was the only official in the U.S. Weather Bureau who recognized and studied the importance of the storm tides in forecasting hurricanes resulting from tropical storms."

But a question haunted him: Did some of the blame for all those deaths in Galveston belong to him? He blamed himself, certainly, for the loss of his wife. His decision to weather the storm in his house had been foolhardy, as had been his advice to some of the people he encountered on Saturday, among them Judson Palmer, who had lost everything. Isaac kept returning to the question. He told and retold the story of how he had asked a reporter for the Associated Press if anything more could have been done to warn the citizens of Galveston—and how the reporter replied, "Nothing more could have been done than was done."

Isaac's subsequent reports to the *Monthly Weather Review* suggest a man obsessed with proclaiming his own prowess at warning of troublesome weather. Unlike his peers, who filed their routine district reports in spare, self-quashing language, Isaac praised his own work, or quoted newspapers and letters that did likewise. In September 1909, for example, he quoted a letter to him praising his warnings of a hurricane that struck Louisiana: "'We feel that your office was solely instrumental in saving to New Orleans, through advices sent out by you in advance, many lives and thousands of dollars worth of property.'"

The Galveston hurricane irrevocably collapsed the wall Isaac had erected between the personal and the professional, the irrational and the rational. On the morning of February 10, 1901, Isaac came forward "on profession of faith" to seek formal admittance to the Baptist church. A month later, the congregation convened at the YMCA pool to conduct its first baptism since the storm. Judson Palmer was there. So were Rosemary and Allie May, and of course Isaac's baby, Esther, and a hundred members of the church. When Isaac stepped into the pool, applause rang for what seemed like hours.

Art became his passion. It filled his spare time with the scent of linseed oil, the seductive texture of canvas. He divided his annual leave into segments as short as two hours so he could attend auctions and estate sales. He collected Early American portraits and Chinese bronzes and in 1918 sold a portion of his collection for the then-fabulous sum of twenty-five thousand dollars. When he had his photograph taken, he knew exactly what he was doing.

He retired in 1935, at the bureau's request, and opened a small art shop on Peter Street in New Orleans. He never remarried. He mourned the passing of slower days before cars and aircraft, but he filled his time to the maximum. He filled it with burnt umber and cerulean blue, linseed oil and turpentine, and the cold caress of ancient bronze.

"Time lost can never be recovered," he said, "and this should be written in flaming letters everywhere."

Isaac Monroe Cline died at 8:30 P.M., August 3, 1955, at the age of ninety-three, just as Hurricane Connie emerged from the Caribbean. Joseph died a week later. The two had not spoken for years.

The Law of Probabilities

WILLIS MOORE BELIEVED the Galveston hurricane to be a freak of nature. "Galveston should take heart," he wrote, "as the chances are that not once in a thousand years would she be so terribly stricken." But another intense hurricane struck in 1915. It hurled a schooner and its crew over the top of the seawall into the city. Throughout the storm, there was dancing at the Hotel Galvez. Other hurricanes struck or came very near in 1919, 1932, 1941, 1943, 1949, 1957, 1961, and 1983. The 1961 storm was Carla, which caused the mass evacuation of a quarter million people from Galveston and surrounding lowlands. The seawall held Carla at bay, but the storm, as if frustrated, launched four tornadoes into the city, destroying 120 buildings over twenty blocks.

The death toll in Galveston from all these hurricanes together was under one hundred, yet toward the end of the twentieth century, meteorologists still considered Galveston one of the most likely targets for the next great hurricane disaster. Unlike their peers in the administration of Willis Moore, they feared that the American public might be placing too much trust in their predictions. People seemed to believe that technology had stripped hurricanes of their power to kill. No hurricane expert endorsed this view. None believed the days of mesoscale death were gone for good. The more they studied hurricanes, the more they realized how little they knew of their origins and the forces that governed their travels. There was talk that warming seas could produce hypercanes twice as powerful as the Galveston hurricane. Insurance companies, appalled by Hurricane Andrew and fearing much worse, quietly began pulling out of vulnerable areas. In the last years of the century a hurricane with the banal name Mitch killed thousands in Latin America and sank a lovely sail-powered passenger ship. The Army Corps of Engineers discovered a curious quirk in the New York–New Jersey coastline and proposed, soberly, that even a moderate hurricane on just the right track could drown commuters in the subway tunnels under Lower Manhattan. The seas rose; summers seemed to warm; the Bering Glacier began to pulse and flow just as it had one hundred years before.

But in the narrow blue-bordered lands of Galveston, extravagant new homes rose on forests of stilts adjacent to blue evacuation signs that marked the island's only exit. Whenever a tropical storm threatened, residents converged on the city's gleaming Wal-Mart to buy batteries and flashlights and bottled water. Once, in a time long past when men believed they could part mountains, a very different building stood in the Wal-Mart's place, and behind its mist-clouded windows ninety-three children who did not know better happily awaited the coming of the sea.

NOTES

It is one thing to write Great Man history, quite another to explore the lives of history's little men. Theodore Roosevelt left volumes of material; Isaac Monroe Cline left little. Indeed, all that Isaac possessed prior to September 8, 1900, was destroyed. How, then, does one fill in the blanks? I approached the problem the way a paleontologist approaches a collection of bones. Even with so little to go on, he manages to stretch over those bones a vision of how the creature looked and behaved. I have been absolutely Calvinist about the bones of this story—dates, times, temperatures, wind speeds, identities, relationships, and so forth. Elsewhere, I used detective work and deduction to try to convey a vivid sense of what Isaac Cline saw, heard, smelled, and experienced in his journey toward and through the great hurricane of 1900.

Luckily, Isaac left a memoir, *Storms, Floods and Sunshine,* published in 1945. It reveals little of his emotional life, but provided insights into the character of late-nineteenth- and early-twentieth-century America that one would be hard-pressed to find elsewhere. Where else could one learn that the state of Arkansas had become so fed up with improper pronunciations of the state's name that it passed legislation making the official pronunciation "Arkansaw"?

I hunted Isaac's trail, too, through the wonderfully rich, achingly fragile archives of the Weather Bureau, lodged in the new National Archives Annex outside Washington—a place that makes deep historical research not a chore but an exciting and always profitable journey.

I touched records, it seemed, that no one had touched for the better part of a century. I handled the very telegrams that Willis Moore, chief of the bureau, himself had touched. I sneezed a lot.

Equally important, if more sterile, were the microfiche copies of Clara Barton's papers at the Library of Congress. Barton knew her place in history. She kept letters and drafts of letters, telegrams and drafts of telegrams, even mundane communications aimed at securing free transportation to and from Galveston. (The Pullman Palace Company gave her a richly appointed Palace car, which railroads agreed to pull at no charge.) Most striking was her growing frustration at the discord that always seemed to accompany her forays into the field.

The single most valuable trove of documents on the hurricane, however, lies in Galveston's Rosenberg Library, God's gift to any student of the great hurricane. The library has hundreds of letters and personal accounts that describe the storm, and over four thousand photographs, some quite macabre. I mined the library's holdings for anything that might provide a fragment of my dinosaur's skin. I used photographs as original documents and spent hours studying them with a magnifying glass. I used details from these photographs to decorate the scenes in *Isaac's Storm*. For example, I describe in one section what Isaac saw from the house where he and his daughters came ashore the night of the storm. Incredibly, the Rosenberg archive has a photograph of exactly that view.

One resource I found exceptionally useful was the library's very detailed map of Galveston in 1899 (see "Fire Insurance Map," in Sources), an immense bound volume that told me Isaac's house was one of the largest in the neighborhood, that it had a slate roof, a small stable out back, and porches or "galleries" on the north and south sides. The map showed me, too, where his house stood in relation to the homes of neighbors like Dr. Samuel O. Young and Judson Palmer. It showed me that as Isaac headed toward the city Saturday morning after his first visit to the beach he would have passed near a wood-planing mill, a bulk coffee roaster, and numerous livery stables, some occupying entire blocks. Each must have perfumed the day. Anyone transported to Isaac's time, I contend, would have found the air permeated by the scent of horse sweat and manure.

In places I relied on my own observations. I did so, for example, in describing the big fat dragonflies of Galveston Island, the behavior of seagulls in a north wind, and the colors of wave crests during a tropical storm. I was lucky enough on one visit to arrive just after a severe tropical storm and before the arrival of another. At one point, as the sun fell, I found myself lost on a narrow spit of land somewhere east of Corpus Christi, with radio newscasters reporting that everyone living near the beach was being urged to evacuate by nightfall.

The sea never looked so lovely, and so deadly.

I HAVE TRIED to keep these notes as concise as possible. Where a citation refers to a document used only once, a full description of the document follows immediately. In all other cases, the citation refers to a more complete bibliographic reference in the Sources section.

Telegram

1. *Do you hear:* Telegram, National Archives. General Correspondence.

The Beach: September 8, 1900

3. Not everyone found Galveston a fairy land of wealth and gleaming streets. The sixteenth-century Spanish called it the Isla de Malhado, the "Isle of Misfortune." Yellow fever scourged the place in 1867 and prompted Amelia Barr, a resident whose husband and two sons died in the outbreak, to call it the "city of dreadful death." On the hottest days, she wrote, the city's tropical climate could be unbearable. "An hour or two of pouring, beating, tropical rain, and then an hour or two of such awful heat and baleful sunshine, as the language . . . has no words to describe." The port thrived, but at the expense of global goodwill. The

Galveston Wharf Company held such monopolistic control over the wharves that the company became known from New York to Liverpool as the Octopus of the Gulf. Gen. P. H. Sheridan took the occasion of a visit to Galveston to issue one of the most infamous geographical slanders of all time. "If I owned Texas and hell," he said, "I'd rent out Texas and live in hell."

3. *He taught Sunday school:* First Baptist Church.

3. *He paid cash:* Giles Mercantile Agency reference book.

4. *"I suppose there is not":* National Archives. Inspection Reports, Galveston, November 1893.

4. *A New Orleans photographer:* Photograph. Isaac Cline. Louisiana State Museum. Whitesell Collection. Accession No. 1981.83.198.

5. *It was a time:* McCullough, *Path*, 247.

5. *She was pregnant:* Isaac Cline, *Monthly Weather Review,* Sept. 1900.

5. *Temperatures in Galveston:* Daily Journal.

6. *For the first time:* "The Incredible Shrinking Glacier."

6. *A correspondent for:* "The Galveston Horror."

6. *In a pamphlet:* In Ousley, Appendix. See also *Immigrants Guide to Western Texas.* Galveston, Harrisburg and San Antonio Railroad. Galveston, 1876. "It is A FABLE, generally believed in the North, that Texas is a land of snakes, tarantulas, scorpions, fleas and mosquitoes . . ." 103.

7. *On Sundays:* Isaac never actually says he and his family visited Murdoch's and the Pagoda on Sundays, but given their proximity to his house, the communal character of the time—and the absence of television—it is all but certain that the Clines did so. Bathhouse details: Fire Insurance Map.

7. *An electric wire: Picturesque Galveston,* 10.

8. *The thudding:* Cline, Isaac, "Special Report," 372.

8. *Isaac woke:* Ibid., 372; also, Cline, *Storms,* 93; Cline, "Cyclones," 13; Cline, "Century," 26; Cline, Joseph, *Heavens,* 49.

8. *Joseph too:* Cline, Joseph, *Heavens,* 49.

9. *For days, however:* Weems, 8–12. But see, especially, National Archives: Weather Bureau. General Correspondence. Sept. 7, 1900, William B. Stockman to Col. H. H. C. Dunwoody, summarizing reports on the storm's early character and track. Box 1475.

9. *The bureau had long banned:* Whitnah, 215.

10. *With most of the block:* Fire Insurance Map.

10. *At the corner:* Photograph. 2502 Avenue Q. Rosenberg Library. Street File: Avenue Q. Also, Fire Insurance Map.

11. *Dr. Samuel O. Young:* Young account, Storm of 1900 Collection. Subject File. Rosenberg Library.

11. *To her, palms and live oak:* Rollfing, 1: 1. Louisa's autobiographical account of her migration to America and her subsequent life would make warm, revelatory reading for any student of the immigrant experience.

12. *In the summer of 1900:* Mason, 54–56.

12. *Josiah: Gregg* Gregg, 101.

12. *The New York* Herald: Eisenhour, 1.

13. *On Friday, September 7, Isaac had read:* In no document does Isaac Cline actually say he read the census report in the Galveston *News,* but it was the biggest local news story of the day. Isaac most certainly read it. See also Weems, 23.

13. *One of its French chefs:* Eisenhour, 4.

13. *"A child's white hearse":* City Directory: Advertisement, J. Levy and Brothers.

14. *The crests of the waves:* Author's observation of how tropical storms influence the surf off Galveston.

14. *Isaac knew the low-pressure center:* Cline, "Address."

14. *"If we had known":* Cline, "Cyclones," 13.

14. *The northeast wind brought:* The locations of the planing mill, bulk coffee roaster, and the many livery stables are set out in the Fire Insurance Map for Galveston.

15. *"Only one-tenth of an inch":* Cline, "Special Report," 372.

15. *"The usual signs":* Ibid., 372.

15. *"Such high water":* Ibid.

15. *Isaac's concern:* Cline, "West India Hurricanes."

PART I: THE LAW OF STORMS

The Storm: Somewhere, a Butterfly

To reconstruct the origins and early travels of the Galveston hurricane, I relied on books and papers by the twentieth century's most significant hurricane researchers, among them William Gray, Christopher Landsea, R. H. Simpson, Richard Anthes,

Kerry Emanuel, and the two Pielkes, Roger junior and senior (see Sources). I found the Pielkes' *Hurricanes,* published in 1997, to be especially useful. My description of how an easterly wave cycle might be perceived by the crew of a ship is based largely on an extended conversation with Hugh E. Willoughby, head of the Hurricane Research Division of the National Oceanic and Atmospheric Administration. I also included, indirectly, the observations of researchers I interviewed in 1998 for a *Time* magazine article on intense hurricanes yet to come, among them Landsea; Gray; Pielke junior; Mark DeMaria, chief of technical support for NOAA's Tropical Prediction Center; Jerry Jarrell, director of the Tropical Prediction Center; and Nicholas K. Coch, self-styled "forensic hurricanologist," Queens College, New York.

20. *Three children died:* Galveston *News,* Aug. 13, 1900.

20. *"The air near the surface":* Garriott, "Forecasts," 321.

20. *Springfield, Illinois, reported:* Galveston *News,* Aug. 13, 1900.

21. *In August, mean temperatures:* Bigelow, "Report," 47, 51, 54, 68, 70, 84, 125, 135.

21. *From mid-July:* Galveston *News,* July 14, July 15, Aug. 10, 1900.

21. *Ten billion joules:* Galveston *News,* Sept. 1, 1900.

21. *Crickets swarmed:* Ibid.

22. *Others became massive:* For excellent descriptions and illustrations of clouds, see *International Cloud Atlas.*

23. *Ships directly in the path:* Author interview, Hugh Willoughby, Hurricane Research Division, National Oceanic and Atmospheric Administration.

25. *"Brown is the new color":* Galveston *News,* Aug. 2, 1900.

25. *Every day an ad:* Galveston *News,* Aug. 1, 1900.

26. *"But suppose":* Zebrowski, 264.

26. *"Could a butterfly":* Ibid., 263.

26. *"One simulated storm":* Ibid., 265.

27. *"Add a little glitch":* Ibid., 266.

Washington, D.C.: Violent Commotions

In this chapter, I relied heavily on the memoirs of Isaac and Joseph Cline, and on two fine histories of weather and the weather service, David Laskin's *Braving the Elements* and Donald Whitnah's *A History of the United States Weather Bureau.*

28. *As a hobby:* Cline, *Storms,* 14–17.
28. *In fall, at acorn time:* Joseph Cline, *Heavens,* 44.
28. *Isaac's uncle swore:* Ibid., 8.
28. *Stories circulated:* Ibid., 10–11. "As I look back on it now," Joseph wrote of the wild man, "it was a soul-sickening spectacle to see a human being, if one could call him that, in such a pitiable plight."
28. *Another turned Boyd's Pond:* Ibid., 13.
29. *The law of convenient epiphany:* Cline, *Storms,* 23–24.
29. *He read everything:* Ibid., 23.
29. *his greatest dream was to write:* Ibid., 23.
29. *"I first studied to be a preacher":* Ibid., 26.
30. *The president of:* Ibid., 27.
30. *A marker showed:* Ibid., 30.
31. *But mainly:* Ibid., 30.
31. *"Meteorology has ever been:* Laskin, 138.
31. *Mark Twain, merciless:* Ibid., 146.
32. *In 1881, police:* Whitnah, 46–47.
32. *Complaints also rose:* Ibid., 46–53.
32. *The assault got personal:* Ibid., 53.
33. *"You will cheerfully":* Frankenfield, 4.
34. *Isaac led them:* Cline, *Storms,* 32.
34. *At Fort Myer, Isaac took apart:* Ibid., 33.
34. *The word* madman: National Archives: Administrative. "Telegraph Cipher." Box 1.
34. *"Paul diction sunk":* Whitnah, 26.
35. *These nocturnal sessions:* Frankenfield, 6.
35. *Often recruits told each other:* von Herrmann, 4.
35. *One lieutenant deliberately:* Geddings, 9.
35. *Another officer, seeking:* Ibid., 3.
35. *One morning a recruit:* Ibid., 11.

The Storm: Monday, August 27, 1900

Here, and in subsequent chapters, I relied on an unpublished report by Jose Fernandez-Partagas, a late-twentieth-century meteorologist who re-created for the National Hurricane Center the tracks of many historical hurricanes, among them the Galveston hurricane. He was a meticulous researcher given to long hours in the library of the University of Miami, where he died on August 25, 1997, in his favorite couch. He had no money, no family, no friends—only hurricanes. The hurricane center claimed his body, had him cremated, and on August 31, 1998, launched his ashes through the drop-port of a P-3 Orion hurricane-hunter into the heart of Hurricane Danielle. His remains entered the atmosphere at 28 N, 74.2 W, about three hundred miles due east of Daytona Beach.

All references to the storm's latitude and longitude in this chapter and subsequent chapters come from page 108 in Fernandez-Partagas's report.

36. *The first formal sighting:* Fernandez-Partagas, 96, note 1.

Fort Myer: What Isaac Knew

37. *He read how men caught:* No one can know precisely what Isaac read at Fort Myer, but it is reasonable to conclude that he studied storms in great depth, since in his time storms were the "meteors" of the greatest interest. For horseflies, see Rosser, The Law of Storms, 40; for stranded deer, see Ludlum, 61; for levitated cannon, Ludlum, 62, 70. In Barbados in 1831 a hurricane carried a 150-pound piece of lead over 1,800 feet, and a 400-pound piece 1,680 feet. McDonough, 7. McDonough found also that the Barbados storm caused a strange change in ambient light. "While this storm was passing over the West Indies . . . objects which were of a whitish color appeared to be of light blue, so marked as to attract the attention of all the inhabitants." McDonough, 8.

37. *Thomas Jefferson kept:* Hughes, 26. George Washington also kept a weather diary, Hughes tells us, and made his last entry the day before his death.

37. *Samuel Rodman Jr.:* Ibid., 31–32.

38. *The first "scientific":* definition Frisinger, 5.

38. *The first person to show:* Ibid., 47.

38. *Aristotle proved:* Ibid., 67.

39. *With the sobriety:* Ibid., 32.

39. *Columbus set off:* Morison, 198; Ludlum, 1–3.

39. *"The weather was":* Morison, 201.

40. *Columbus and his captains:* Ludlum, 1; Douglas, 47.

42. *Columbus had at least:* Morison, 584–93.

42. *"The storm was terrible":* Ludlum, 6.

43. *Only one ship:* Morison, 590.

43. *In 1638, Galileo:* Frisinger, 67.

43. *The meteorological significance:* Ibid., 68.

44. *To get any observable effect:* Ibid., 68–69.

44. *The term* barometer: Ibid., 72.

44. *Isaac considered him:* Cline, "Century," 3.

45. *The* Reserve *put in:* Snow, 1–17; Douglas, 132–35.

48. *In 1627, a very brave:* Lockhart, 37.

49. *It was Edmund Halley:* Frisinger, 123–25.

49. *In 1735, George Hadley:* Ibid., 125–28.

50. *A century later:* Lockhart, 37; Watson, 28; Trefil, 96–104.

50. *Isaac, in his 1891 talk:* Cline, "Address."

51. *"Who can attempt":* Thomas, 164. Thomas reproduces Archer's full account on 154–69. See also Reid, 296–303. For detailed accounts of the three hurricanes, see Reid, Douglas, and Ludlum. For official death toll, see Rappaport and Fernandez-Partagas, "Deadliest."

51. *The second hurricane:* Ludlum, 69–72; Reid, 340–65.

51. *"The most beautiful island":* Ludlum, 69.

51. *The storm lurched:* Reid, 345.

51. *The third hurrican:* Douglas, 173; Ludlum, 72–73.

51. *Together the three:* Douglas, 173.

52. *The first captain:* Friendly, 146.

52. *On September 3, 1821:* Ludlum, 81; for complete account, see 81–86. Also, Douglas, 221–26; Rosser, W. H., 9–17. For an account of the long-burning feud between Redfield and James Espy, "The Storm King," see Laskin, 138–40.

53. *It was Piddington:* Douglas, 224.

53. *"The unfortunate inhabitants":* Tannehill, 31.

54. *"I had studied":* Cline, "Century," 25–26.

54. *Piddington, in his immensely popular:* For example, Piddington, 102–3 and 134–37.

54. *"you have the hurricane in your hand":* Piddington, 134.

54. *As one nineteenth-century captain:* Rosser, W. H., 41.

55. *The chief did not want:* For grave robbing, see Whitnah, 75. For sex scandals, see Cline, *Storms,* 76–78.

55. *Isaac gave a beauty queen's answer:* Cline, ibid., 35.

55. *"I was twenty-one":* Ibid.

The Storm: Tuesday, August 28, 1900

56. *The earth's rotation:* Author interview, Willoughby (see note for page 23). Cyclones in the northern hemisphere always rotate counterclockwise, in the southern hemisphere, clockwise, which helps explain why no hurricane can cross the equator. I simply could not understand how a counterclockwise cyclone could be generated by right-veering Coriolis winds, until Willoughby patiently explained the process. Blame for the pool-cue analogy, however, belongs entirely to me. See also Tannehill, 5–6.

56. *On Tuesday, August 28:* Fernandez-Partagas, 96, note 1.

Galveston: Dirty Weather

57. *"Something new":* Cline, *Storms,* 39.

57. *"They evidently learned":* Ibid., 39.

58. *It did exist:* Ibid., 45.

58. *"I was told":* Ibid., 45.

59. *"That looks like":* Ibid., 46–47.

60. *Far to the north:* McCullough, *Mornings,* 316–37.

60. *Roosevelt called:* Ibid., 337.

61. *During a visit:* Cline, "Summer," 336.

62. *It was August:* Cline, *Storms,* 51.

62. *He tracked down reports:* Ibid., 52.

63. *"She was a beautiful":* Ibid., 57.

63. *An inefficient man:* Traxel, 42.

64. *"In the past":* Ibid., 42.

64. *In Greely's first year:* Cline, *Storms,* 65.

64. *He fired:* Ibid., 65.

64. *A fondness for:* Ibid., 66.

65. *An observer in the Midwest:* Ibid., 66.

65. *On January 21, 1888:* National Archives: Inspection Reports, Galveston, January 1888.

65. *And then came:* Laskin, 146–47.

67. *The city had:* Turner, Elizabeth, 24.

67. *Through the Negro Longshoremen's Association:* Mason, 51. See also Turner, Elizabeth, 371–72.

68. *On November 13, 1893:* National Archives: Inspection Reports, Galveston, November 13–15, 1893.

69. *At the time of Harrington's appointment:* Cline, *Storms,* 74.

71. *In a later memoir:* See Cline, Joseph, *Heavens.*

71. *Morton's assault:* Abbe, Container 8.

71. *"Nearly every real advance":* Ibid., June 17, 1893.

72. *"Dunwoody is a selfish intriguer":* Whitnah, 79.

73. *The system, he told Congress:* Ibid., 87.

73. *Moore, greedy for:* National Archives: General Correspondence. Letter, March 29, 1900, Willis Moore to Owen P. Kellar.

73. *"I knew," Moore wrote:* Tannehill, 110–12.

74. *"Wilson," he said:* Ibid., 112.

74. *He told Moore:* Ibid., 112.

75. *At breakfast:* Cline, *Storms,* 83.

75. *Indeed, in that same hurricane season:* Miller, x (Introduction).

76. *On November 26:* Traxel, 296–99.

76. *The sudden cold:* Cline, *Storms,* 88.

The Storm: Thursday, August 30, 1900

78. *At 9:00 A.M.:* Alexander, 380.

78. *"About 10 P.M.":* Ibid., 380.

78. *There was, according to the Antigua:* Standard Ibid., 380.

Galveston: An Absurd Delusion

79. *In January 1900:* Coulter, 63.

80. *He explained first:* Cline, "West India Hurricanes."

80. *"No greater damage":* Ibid.

80. *"The damage from the storm":* Ibid.

81. *"The single tornado":* Ibid.

81. *By 5:00 P.M.:* Greely, 444.

81. *"This evidence of"*: Ibid., 444.

82. *"ebb surge"*: I first heard this term from Nicholas K. Coch, professor, School of Earth and Environmental Science, Queens College, City University of New York.

82. *"The tide now swept"*: Greely, 444.

82. *But Gen. Adolphus Greely:* Ibid., 443.

83. *"The water in the bay"*: Tannehill, 35.

83. *"The appearance of the town"*: Ibid., 35.

83. *the Progressive Association:* Mason, 74.

84. *The city's engineer:* Ibid., 74.

84. *The city's* Evening Tribune: Ibid., 74

84. *"But," engineer Hartrick wrote:* Ibid., 74.

84. *"It would be impossible"*: Cline, "West India Hurricanes."

PART II: THE SERPENT'S COIL

The Storm: Spiderwebs and Ice

87. *The storm entered:* Fernandez-Partagas, 108.

87. *As vapor rose:* For an excellent discussion of clouds and cloud physics, see Volland.

88. *But hurricanes defeat this cycle:* For a comprehensive analysis of this phenomenon, see Liu.

88. *In 1979 a tropical storm:* Henry et al., 22.

88. *A Philippine typhoon:* Ibid., 22.

88. *Ninety-six and a half inches:* Tannehill, 72.

88. *Hurricane Camille:* Pielke, Roger A., Sr., 2–3.

89. *Camille's rain fell:* Ibid., 3. Hugh Willoughby of the Hurricane Research Division, in reading the manuscript of *Isaac's Storm* for accuracy, called this an urban legend.

Galveston: Louisa Rollfing

I based this entire chapter on Louisa Rollfing's autobiography, in the Galveston Collection of the Rosenberg Library.

The Levy Building: Isaac's Map

98. *At three o'clock:* Galveston *News,* Sept. 5, 1900.

98. *At the police station:* Ibid.

98. *Isaac heard the first clap:* Daily Journal.

99. *One of the newest arrivals: The New York Times,* Sept. 11, 1900, 3. (See "Vessels at Galveston.")

99. *Isaac sent a man:* Young, 1.

99. *Throughout July:* National Archives: General Correspondence. Letters: James Berry to Official in Charge, Galveston, July 5, 1900; Isaac Cline to Weather Bureau, July 9, 1900; James Berry to Official in Charge, Galveston, Aug. 16, 1900; Isaac Cline to Weather Bureau, Aug. 19, 1900. Box 1423.

100. *He told Secretary:* National Archives: Letters Sent. Moore to Wilson, Sept. 15, 1900.

100. *Baldwin left:* National Archives: Administrative. Box 7. Slip Book. Aug. 29, 1899–Oct. 23, 1900. No. 425.

100. *Moore promised:* National Archives: General Correspondence. Telegram, Aug. 20, 1900. Box 1473. See also letters (No. L.R.7510-1900): Acting Chief Clerk to Official in Charge, New York, and Acting Chief Clerk to Official in Charge, Galveston, both of Aug. 22, 1900.

100. *He telegraphed Moore:* National Archives: General Correspondence. Telegram, Aug. 20, 1900.

100. *For the last week, Young:* Young, 1.

101. *"He agreed with me":* Ibid.

Cuba: Suspicion

102. *Through Dunwoody, Moore persuaded:* National Archives: General Correspondence. Letter, Moore (as acting secretary of agriculture) to Gen. T. T. Eckert, Western Union, Aug. 28, 1900. Box 1475.

102. *Cuba's meteorologists had pioneered:* Douglas, 230–36; Hughes, 13; Tannehill, 63.

103. *"It was at first very difficult":* National Archives: Records of Surface Land Observations. Records Relating to Hurricane Display Systems in the West Indies. Report, Fiscal Year Ending June 30, 1899. William B. Stockman. Box 1.

104. *Internal memos flew:* National Archives: General Correspondence. Box 1470. The saga begins with Stockman's report of July 31, 1900. At one point, Moore's

office advises Stockman, "it is thought well to suggest care on your part that you may not allow your work to get beyond your strength; and thus impair, through physical disability, the excellent record made by you at Havana." Garriott's wonderfully crafted note is numbered L.R. 7057-1900. Moore's note is scrawled in pencil at the bottom of a memo slip in the same file.

104. *It was paramount:* National Archives: General Correspondence. Letter, July 6, 1900, Stockman to Official in Charge, St. Kitts.

104. *He spent a good part:* National Archives: General Correspondence. The complete story lies in Box 1471. It begins with a letter from a secret informant to Stockman, dated Aug. 8, 1900, and ends with Moore's terse letter of Sept. 6, 1900, two days before the Galveston storm.

105. *On August 24, 1900, W. T. Blythe:* National Archives: General Correspondence. Letter, Aug. 24, 1900, Blythe to Moore. Box 1475.

105. *On August 28, Willis Moore:* National Archives: General Correspondence. Letter, Moore (as acting secretary of agriculture) to Gen. T. T. Eckert, Western Union, Aug. 28, 1900. Box 1475.

106. *"This conduct":* National Archives: General Correspondence. See clipping, Sept. 2, 1900, in Box 1475.

106. *"A very bitter opposition":* National Archives: General Correspondence. Letter, Sept. 5, 1900, H. H. C. Dunwoody to William Stockman. Box 1475.

107. *On Saturday, September 1:* National Archives: General Correspondence. See clipping from *La Lucha*, Sept. 1, 1900. Box 1475.

107. *On August 31, Julio Jover:* National Archives: General Correspondence. See clipping from *La Lucha*, Sept. 3, 1900, containing Jover's dispatch of August 31. Box 1475.

107. *The next day, Belen's Father Gangoite:* National Archives: General Correspondence. See clipping from *Diario de la Marina,* Sept. 2, 1900, containing Gangoite's dispatch of Sept. 1, 1900. Box 1475.

108. *By Friday, the total:* Monthly Weather Review, Sept. 1900, 377.

New Orleans: Captain Halsey's Choice

109. *At 9:20 A.M. Wednesday:* Monthly Weather Review, Sept. 1900, 374; Fernandez-Partagas, 101, note 34; *The New York Times,* Sept. 11, 1900, 3.

109. *Wrote Piddington:* Piddington, 376–77.

110. *The* Louisiana *entered:* Fernandez-Partagas, 101, note 34.

Straits of Florida: A Matter of Divination

111. *Shortly after noon:* The Daily Register, Mobile, Ala., Sept. 6, 1900.

111. *"We are today near":* National Archives: General Correspondence. See clipping from *La Lucha*, Sept. 6, 1900, quoting Jover's dispatch of 8 A.M. Sept. 5. Box 1475.

112. *Winds reached:* Fernandez-Partagas, 99, note 21.

112. *In Key West:* Ibid., 99, note 22.

112. *Its velocity dropped:* Ibid., 99, note 21.

113. *The next morning:* National Archives: General Correspondence. See clipping from Havana *Post*, Sept. 7, 1900, quoting Stockman's dispatch. Box 1475.

113. *Two hours later:* Letter, E. M. Vernon, chief, Forecasts and Synoptic Reports Division, to M. S. Douglas, Nov. 9, 1956. Vernon, in response to an inquiry from Douglas, apparently for her book, *Hurricane,* wrote, "We can find no reference to the issuance of hurricane warnings for the Texas or Louisiana coasts for this hurricane." Rosenberg Library. 95-00020. Box 1, File 7.

113. *"Advise quick":* National Archives: General Correspondence. See telegram, Sept. 6, 1900, from Ocean Fishery, Long Branch, N.J., to Weather Bureau, Washington. Box 1475.

114. *"Not safe to leave":* National Archives: General Correspondence. Telegram, Sept. 6, 1900, Chief Willis Moore to Ocean Fishery, Long Branch, N.J.

114. *He told Jover:* National Archives: General Correspondence. See clipping from *La Discusion*, Sept. 11, 1900, and attached translation, containing interview with Dunwoody. Box 1475. At one point Jover exclaims, ". . . I believe that nobody has the right to forbid a citizen telegraphing to a newspaper all that he wishes, be it true or false." To which Dunwoody responds, "Well I understand that it is not just but can not the government do what it pleases? Moreover the government has a meteorological Bureau and it does not need any more."

Key West: M Is for Missing

115. *The map that reached Erie:* National Archives: General Correspondence. National Weather Map, Erie, Pa., Sept. 6, 1900. Box 1475.

Gulf of Mexico: The Devil's Voice

117. *Once past the bar:* Fernandez-Partagas, 101, note 34.

117. *At 6:00 A.M. Thursday:* The New York Times, Sept. 11, 1900.

118. *At one o'clock, Halsey:* Ibid.

118. *"I do not like to speak:* Ibid.

118. *The* Louisiana *rose clear:* Ibid.

120. *In 1912, the Reverend J. J. Williams:* Tannehill, 18.

120. *The frightened Malay:* Piddington, 208.

120. *To Gilbert McQueen:* Reid, 92.

120. *One of the strangest:* Reid, 73–76; also Piddington, 340.

122. *On September 1, 1923:* Tannehill, 128.

122. *A Weather Bureau meteorologist:* Ibid., 129.

122. *In Galveston, Thursday:* Daily Journal.

The Storm: Swells

124. *The tallest wave:* Lockhart, 115.

125. *A tsunami:* Zebrowski, 134. Zebrowski tells the story of the U.S.S. *Wateree,* a paddle steamer caught in a tsunami that came ashore in northern Chile on Aug. 13, 1868. Investigators discovered the wave continued traveling another 5,580 miles to strike the Sandwich Islands twelve hours and thirty-seven minutes later. They computed an average speed of five hundred miles per hour. Rear Admiral L. G. Billings wrote, "Looking seaward, we saw, first, a thin line of phosphorescent light, which loomed higher and higher until it seemed to touch the sky; its crest, crowned with the death light of phosphorescent glow, showing the sullen masses of water below." The *Wateree* landed upright and intact three kilometers inland. The U.S. Coast and Geodetic Survey later estimated the tsunami had risen to seventy feet in height. See Zebrowski, 131–35.

Galveston: Heat

126. *He was a veteran:* Galveston *News,* Sept. 13, 1900; see also Weems, 20–22, 26–27, 46. For details of ship's size and ownership, see "Vessels at Galveston" in *The New York Times*, Sept. 11, 1900. Also, in the Rosenberg Library's vast collection of storm photographs I found a post-storm photograph of a barge being loaded with corpses. A large ship is moored near one end of the barge. Close examination with a magnifying glass shows the ship is the *Pensacola.* Members of her crew stand at the bow watching the macabre proceeding. Rosenberg Library. Storm of 1900 Collection. Photograph G-1771, File 1.2. No. 9.

126. *Simmons pulled out:* Weems, 20–21.

127. *He did note, however:* Ibid., 21.

127. *At 9:35 A.M. Galveston time:* Daily Journal.

127. *"Well, young man":* The New York Times, Sept. 23, 1900.

128. *"Menard," Simmons said:* The New York Times, Sept. 13, 1900.

129. *"It looked as if ":* Ibid.

130. *There would be no:* Cartwright, *Galveston,* 167.

130. *"Thursday afternoon":* Young, 1.

131. *"For my own satisfaction":* Ibid., 1.

131. *"The error I made":* Ibid., 1.

131. *That evening, at precisely:* Observations.

Cuba: "Who Is Right?"

132. *Dunwoody had written:* National Archives: General Correspondence. Letter, Sept. 5, 1900, H. H. C. Dunwoody to William Stockman. Box 1475.

133. *"No dangerous winds":* National Archives: General Correspondence. Letter, William Stockman to H. H. C. Dunwoody, Sept. 7, 1900, page 14. Box 1475.

133. *Any comparison of:* Ibid., 15.

134. *"At day-break":* National Archives: General Correspondence. See Father Gangoite's dispatch in clipping from *La Lucha,* Sept. 10, 1900. Box 1475.

PART III: SPECTACLE

Observation

137. *"The sky seemed":* Mason, 78–79.

Gulf of Mexico: The Pensacola

138. *At 10:30 that morning:* Galveston *News,* Sept. 13, 1900.

The Beach: Delight

140. *The other was:* Young, 1.

140. *"I was certain":* Ibid., 1–2.

140. *Later Isaac took:* Cline, *Storms,* 93.

141. *If not for him:* Letter, E. M. Vernon, chief, Forecasts and Synoptic Reports Division, to M. S. Douglas, Nov. 9, 1956. Rosenberg Library. 95-00020. Box 1, File 7.

141. *Bornkessell replied:* Weems, 45.

141. *"The storm was":* Personal Accounts: Blagden, 6.

141. *He advised them:* Cline, "Special Report," 373.

141. *One resident, Sarah:* Personal Accounts: Hawley, Sarah, 1.

142. *The car was crowded:* Personal Accounts: Goodman, 2.

142. *On Sunday, September 2:* Galveston *News,* Sept. 2, 1900.

142. *They found:* Galveston *News,* Sept. 8, 1900.

144. *"There have been":* Rosser, Angie.

145. *Many decades later:* Ibid.

145. *"Even so":* Personal Accounts: Cortes, 2.

145. *That morning Mrs. Charles Vidor:* Weems, 37–38.

146. *"I remember now":* Ibid.

146. *"I left home":* Personal Accounts: Hopkins, "The Day," 1. See also other Hopkins accounts.

148. *One witness reported:* Mason, 81.

148. *"The sight was grand":* Personal Accounts: Davis, 2.

150. *"My family pleaded":* Personal Accounts: Wolfram, 1.

150. *"We have had storms before":* Mason, 79.

151. *Judson Palmer, secretary:* First Baptist Church, 1.

152. *"For a while":* Rollfing, 3: 1.

Ritter's Café: *"You Can't Frighten Me"*

155. *Rabbi Henry Cohen:* Nathan and Cohen, 132–45. See also Cartwright, *Galveston,* 145–46, 165–66, and "Blow," 114.

158. *Saturday morning:* Mason, 108–9; see also Personal Accounts: Focke, 4. A friend of Stanley Spencers encountered Mrs. Spencer on Monday, Sept. 10, and learned from her that Mr. Spencer's face seemed wholly unmarked even though the back of his head had been crushed. Personal Accounts: Hawley, J. H., 3–4.

Bolivar Point: The Lost Train

160. *"When we crossed":* Coulter, 95–99; First Baptist, 3; Mason, 84–85.

163. *Another passenger:* Personal Accounts: Benjamin.

163. *Poe lived in Lake Charles:* Coulter, 89–90; Mason, 85–86, 162.

166. *Marie Berryman Lang:* Personal Accounts: Lang. See also Weems, 42.

25th and Q: A Gathering of Toads

167. *"The storm swells":* Cline, *Storms,* 93.

168. *By 2:30 P.M., Galveston time:* Ibid., 94.

168. *Isaac reported:* Ibid., 94.

169. *Joseph gave himself:* Cline, Joseph, 51.

170. *Along the way:* Tapp, 8.

171. *Stay put, Isaac said.:* First Baptist Church, 2. Weems, 74.

171. *"Those who lived":* Cline, "Special Report," 373.

171. *"Many went to his house":* Personal Accounts: Blagden, 6.

171. *"Every little board":* Personal Accounts: Bettencourt, 16.

171. *The water was:* Cline, Joseph, 53.

171. *Neck deep:* Cline, "Special Report," 373.

171. *"He knew better":* Cline, Joseph, 53.

172. *Evacuate:* Ibid., 53; Cline, "Special Report," 373.

172. *Stay:* Cline, "Special Report," 373.

PART IV: CATACLYSM

Telegram

175. *We have been:* Telegram. National Archives: General Correspondence.

The East Side: Louisa Rollfing

176. *August Rollfing:* Rollfing, 3: 4.

178. *At about two:* Cline, *Tropical Cyclones,* 246.

178. *She watched quietly:* Rollfing, 3: 3.

Avenue P½: Parents and Their Choices

179. *At two o'clock:* Young, 2.

180. *The water moved fastest:* Author's analysis. Most of the city's streets were lined with high curbs, which acted like erosion gullies to channel the flow of water.

180. *Young saw:* Young, 2. Specifically, Young recalls seeing "wrecked shanties, boxes, barrels, wooden cisterns and everything else that fell in [the current's] power." That carriages and outhouses and myriad other things floated within is beyond question. He makes no mention of seeing bodies, however—although by that time there were many embedded in the current coursing through the city.

181. *"Being entirely alone":* Young, 2.

181. *As Louise Hopkins:* Personal Accounts: Hopkins, "The Day," 1–4.

181. *There were boxes and boards:* Again, the flow carried all manner of debris. It also carried snakes. After the storm, one captain reported encountering snakes far out in the Gulf. See note for page 202, *Venomous snakes.*

182. *Louise noticed that:* Personal Accounts: Hopkins, "The Day," 5.

183. *At precisely 2:30 P.M.:* Daily Journal.

183. *At 5:15, the wind:* Daily Journal.

183. *"We had a warm feeling":* Personal Accounts: Hopkins, "The Day," 6.

184. *"When it was lighted":* Personal Accounts: Hopkins, Interview, 14–15.

184. *A neighbor couple:* First Baptist Church, 2.

185. *It grew cysts:* Personal Accounts: Cortes, 4.

185. *At 7:00 P.M.:* First Baptist Church, 2.

185. *"I cannot pray":* Ibid., 2.

185. *Garry Burnett recommended:* Ibid., 2.

25th and Q: Isaac Cline

188. *"At this time":* Cline, "Special Report," 373.

188. *At 6:30 P.M.:* Ibid., 373.

188. *The sea was strangely flat:* Young, 3.

188. *The Neville house:* Photograph. 2502 Avenue Q. Rosenberg Library. Street File: Avenue Q.

188. *It contained homes:* Rollfing, 3: 7. For the most vivid record of destruction, see the Rosenberg Library's collection of aftermath photographs, some of which are quite grisly. One photograph, G-1771, File 7.5, No. 13, shows a vast plain of

wreckage where Isaac Cline's neighborhood had once existed. Another, G-1771, File 1.2, No. 8, shows six men apparently about to bury the body of a woman. She is well and elaborately dressed in pantaloons with horizontal stripes and a dress printed with bold, broad vertical stripes. The men represent a cross-section of society. One is black, his face deeply furrowed with an expression of distaste. It is possible he is expressing disdain for the photographer. Another member of the party is young, clean-shaven, and handsome, dressed neatly in a long-sleeved white shirt, straw hat, and suit pants. Close examination with a magnifying glass reveals that his eyes are closed. The men stand in a great sea of shattered boards.

189. *"I was standing"*: Cline, "Special Report," 373.

189. *(Joseph claims . . .:* Cline, Joseph, 53.

189. *The brothers herded:* Ibid.

189. *"These observations"*: Cline, "Special Report," 373.

190. *One block north:* Young, 2.

190. *"The debris fairly flew"*: Ibid., 2.

190. *One witness:* Galveston *Galveston News,* Sept. 13, 1900, 5.

190. *Dr. Cline's house:* Author's analysis, based on Young's proximity to Isaac Cline's house, and the orientation of his house. Young, 2. See also Fire Insurance Map. Regarding the absence of Cline's galleries, see Cline, Joseph, 53.

191. *"Strangely enough"*: Cline, Joseph, 53.

192. *"I urged them"*: Ibid., 53.

192. *In Dallas, three hundred:* Acheson, 211; Mason, 106.

192. *(whose Galveston agent:* See death list, Galveston *News,* Sept. 14, 1900. W. Pilford of the Mexican Cable Company is listed, along with his four children, Madge, Willie, Jack, and Georgianna. The list cites the location where they were killed as "Twenty-fifth and Q." Isaac's corner.

192. *At that moment:* Acheson, 211; Mason, 106.

The Levy Building: Vital Signs

193. *Saturday evening, John Blagden:* Personal Accounts: Blagden, 1–2.

193. *Meteorologists discovered:* Cline, "Special Report," 374.

194. *Barometric pressure had fallen:* Daily Journal.

194. *In Galveston harbor:* Weems, 101.

195. *The hull was built:* Ibid., 102.

195. *In the train station:* First Baptist Church, 4.

195. *Years later, scientists:* Rappaport and Fernandez-Partagas, 9.

195. *"Assuming that the reading":* Garriott, "West Indian," 392.

195. *The bureau later estimated:* Monthly Weather Review, Sept. 1900, 424.

195. *Each would generate:* For an excellent discussion of wind force and effects, see Zebrowski, 248–51

195. *Captain Storms:* Mason, 160.

196. *One man tied his shoes:* Personal Accounts: Wolfram, 1.

196. *A survivor identified:* Personal Accounts: "Charlie," 1–2.

196. *One of the deadliest:* Pielke and Pielke, 199.

196. *In 1876 Henry Blanford:* Monthly Weather Review, "What Is a Storm Wave?" Oct. 1901, 461.

197. *In October:* Garriott, "West Indian," 391.

197. *If a hurricane strikes:* Henry et al., 19.

197. *In effect, the storm's trajectory:* Cline, "Relation," 208. Garriott, "West Indian," 391; for a summary of the hurricane's character and path, see 384–92.

198. *The first shift:* Cline, Tropical Cyclones, 246.

198. *At 7:30 P.M., the wind:* Cline, "Special Report," 373. Cline, "Relation," 207.

Avenue P½: The Wind and Dr. Young

199. *About seven o'clock:* Young, 2.

200. *A single cubic yard:* Cline, "Relation," 203.

200. *One man reported:* First Baptist Church, 34.

200. *"It turned partly":* Young, 3.

201. *"The wind at 125:"* Ibid.

201. *"The drops of rain":* Ibid.

202. *Venomous snakes:* Henry et al., 19, 23. Here I make the assumption that phenomena common in later hurricanes were likely to have occurred in the Galveston hurricane. Henry et al. cite Hurricane Audrey, which struck the Louisiana coast in 1957. "It is thought that the majority of people who drowned sought safety by climbing into high trees and then fell into the rising flood waters after they were bitten by snakes also taking refuge in the trees" (19). Later, they state, "Snakes, which are strong swimmers, will be along roads, in the remains of buildings, in trees, and in other high and dry places" (23). A schooner, *Viva,* out of Corpus Christi, arrived in Galveston soon after the storm. A passenger,

Leopold Morris, told a Galveston *News* reporter he saw large snakes swimming in the Gulf. "The snakes seemed to plead for a ride on the boat, and if ever I saw a serpent look kindly toward a human being those were the ones." Galveston *News,* Sept. 16, 1900.

202. *A rocket of timber:* Cartwright, "Big Blow," 114.

202. *At the expensive Lucas Terrace:* Mason, 126–27. For a compelling photograph of Lucas Terrace before and after the storm, see Weems, plates, between 84 and 85.

202. *At another address:* Mason, 157–58.

203. *"The house rose":* Young, 3.

25th and Q: What Joseph Saw

204. *"As the house capsized":* Cline, Joseph, 54.

204. *"All the other occupants":* Ibid.

205. *"I had hoped":* Ibid., 55.

The Beach: Ruby Credo

206. *As soon as Ruby Credo's:* Tapp, 8–9.

207. *"The water was rising":* Ibid., 9.

207. *"When our house":* Mason, 111.

208. *"We could lie back":* Tapp, 10.

25th and Q: What Isaac Did

210. *When the trestle struck:* Cline, *Storms,* 96.

The Beach: A Light in the Window

211. *Her sister, Lois:* Personal Accounts: Hopkins, "The Day," 7.

212. *The ten sisters:* St. Mary's. See also Cartwright, "Big Blow," 115, and Mason, 148–51.

212. *A few older children:* St. Mary's. There is some disagreement as to the name of one of the three survivors. Mason, at 151, identifies him as Francis Bulnavic. The St. Mary's booklet says his name was Frank Madera. I chose Bulnavic arbitrarily. Where all accounts agree is that only three boys among the orphanage's ninety-three children survived.

213. *Later, a rescuer:* St. Mary's. Also, Ousley, 114–15. In another macabre sighting, a steward aboard the Mallory liner *Alamo* said, "One of the saddest sights I saw was the dead bodies of a Sister of Charity and three little boys lashed together floating in the water." *The New York Times,* Sept. 23, 1900.

213. *August Rollfing sat alone:* For August's story, see Rollfing, 3: 4–7.

25th and Q: Isaac's Voyage

217. *He was alone:* Cline, *Storms,* 96.
218. *"My heart suddenly":* Cline, Joseph, 55.
218. *"We placed the children":* Cline, *Storms,* 97.
218. *"Our little group":* Cline, Joseph, 56.
218. *"Sometimes the blows":* Cline, *Storms,* 97.
218. *"At one point":* Cline, Joseph, 57.
219. *They drifted:* Cline, *Storms,* 96–97; Cline, Joseph, 56–57.
219. *A rocket of timber:* Cline, Joseph, 58.
219. *Joseph saw a small girl:* Cline, Joseph, 58; Cline, *Storms,* 97.
219. *"Papa! Papa!":* Cline, Joseph, 59.
219. *And there was this:* Ibid., 57.

PART V: STRANGE NEWS

Telegram

223. *First news from:* National Archives: General Correspondence.

Gulf of Mexico: First Glimpse

224. *About dawn:* Galveston *News,* Sept. 13, 1900.
224. *"We found":* Ibid.

Galveston: Silence

226. *Saturday evening someone:* New Orleans *Daily Picayune,* Sept. 9, 1900.
226. *On Sunday a small party:* Weems, 39.
227. *As one of the region's:* Personal Accounts: Sterett, 1–3. See also Acheson, 205–17.
227. *The swollen bodies:* Personal Accounts: Sterett, 2.
227. *For Sterett:* Ibid.

227. *"And so help me":* Ibid.

228. *"Everything, it seemed":* Ibid.

229. *"It must have taken":* Personal Accounts: Monagan.

229. *"I am an old soldier":* "Galveston Horror," 33.

229. *At one point Sterett:* McComb, 127.

229. *It was a night:* Personal Accounts: Monagan.

229. *They stopped a man:* Ibid.

229. *"Surely the man":* Ibid.

28th and P: Searching

230. *It was, he said:* Cline, "Century," 31.

231. *In the wreckage:* Isaac Cline never says exactly what he saw that morning, but, as hundreds of photographs in the Rosenberg Library storm collection show, there can be no doubt that Isaac saw hats, clothing, corpses, and far more.

231. *One hundred corpses:* Ousley, 120.

231. *Some had double-puncture wounds:* See note for page 202, *Venomous snakes.*

231. *Forty-three bodies:* Ousley, 120.

231. *"There were so many":* Personal Accounts: Tipp, 9

232. *Isaac checked:* What Isaac Cline did in the days immediately after the storm is a mystery. I have based this paragraph and others that follow on my sense of Isaac's character, and on my understanding, derived from scores of personal accounts, of how people throughout Galveston behaved after the storm. That he visited the hospitals and morgues seems beyond question.

232. *J. H. Hawley:* Personal Accounts: Hawley, J. H., 1–2.

232. *A photograph survives.:* Photograph of morgue. Rosenberg Library. Storm of 1900 Collection. G-1771. Folder 1.2. "Bodies." No. 2.

232. *Isaac, moving systematically:* See note for page 232, *Isaac checked.*

233. *Sunday he gave:* Galveston *News*, Sept. 9, 1900.

233. *That morning Father James Kirwin:* Ousley, 116.

233. *Anthony Credo learned:* Tapp, 12.

233. *Soon after Ruby Credo:* Ibid., 10.

233. *Judson Palmer lay:* First Baptist Church, 3.

234. *Later a colleague:* Personal Accounts: Lewis.

234. *John W. Harris was seven:* Personal Accounts: Harris, 7–8.

234. *People moved as if dazed:* Coulter, 224.

234. *"You will hear"*: Ousley, 120.

235. *"Oh God"*: Personal Accounts: Hopkins, Interview, 10.

235. *The storm, Halsey told:* The New York Times, Sept. 11, 1900.

235. *A photograph exists:* Photograph. 27th St. and Ave. N Looking S.E. Rosenberg Library. Storm of 1900 Collection. G-1771. File 7.5. No. 13.

236. *The Muats had expected:* Muat, Thomas. Untitled news clipping. Rosenberg Library. Storm of 1900 Collection. Subject File. News Clippings.

Daily Journal: Tuesday, Sept. 11

238. *I. M. Cline:* Daily Journal.

Galveston: "Not Dead"

239. *The* Tribune *ran:* Galveston Tribune-Post, Sept. 12, 1900.

239. *Soldiers rounded up:* Ousley, 117.

239. *The barge was moored:* Photograph of barge. Rosenberg Library. Storm of 1900 Collection. Photograph G-1771, File 1.2. No. 9.

240. *"It was realized"*: Ousley, 266.

240. *Phillip Gordie Tipp's crew:* Personal Accounts: Tipp, 10.

240. *The city's lifesaving squad:* Coulter, 199.

241. *"The stench from"*: Personal Accounts: Deer, 2.

241. *Emma Beal was ten:* Personal Accounts: Beal, Part I, 9; Part II, 9–10.

241. *One survivor:* Personal Accounts: Stuart, 53.

241. *There was talk:* Galveston News, Sept. 12, 1900.

241. *On Sunday night:* Galveston News, Sept. 17, 1900.

241. *And for William Marsh Rice:* Morris, 84–112.

242. *"Diligent inquiry"*: Personal Accounts: Stuart, 53.

242. *"I do not know"*: Personal Accounts: Blagden, 5.

242. *"Fearful hot"*: Personal Accounts: "Charlie," 5.

242. *"Every day the stench"*: Tapp, 12.

244. *Clara Barton arrived:* Barton. Telegram, Barton to William Howard. Sept. 18, 1900.

244. *She came with a trainload:* Barton. Letter, Barton to Mayor of Galveston. Sept. 20, 1900.

244. *Hearst . . . gave $50,000:* Report, 2.

244. *The Kansas State Insane Asylum:* Barton. Letter, Sept. 25, 1900.

244. *Colored Eureka Brass Band:* Barton. Letter, Sept. 26, 1900.

245. *Elgin Milkine Company:* Barton. Letter, Sept. 19, 1900.

245. *Fraternal Mystic Circle:* Barton. Letter, Sept. 20, 1900.

245. *Ladies of the Maccabees:* Barton. Letter, Oct. 13, 1900.

245. *The city of Liverpool:* Report, 71.

245. *Cotton Association of Liverpool:* Ibid.

245. *New York sent the most:* Barton. Report of Red Cross Relief, Galveston, Texas, 77.

245. *New Hampshire sent:* Ibid.

245. *"It would not surprise me":* Barton. Letter, Oct. 14, 1900.

245. *Among the contributions:* Barton. Letter. Cambria Steel Company to Clara Barton, Sept. 21, 1900; Barton to Cambria Steel Company, Sept. 25, 1900.

245. *Observers within:* National Archives: General Correspondence. Letter, Isaac Cline to chief of Weather Bureau, Nov. 8, 1900. Isaac wrote, "We fail to find language which will express our feelings of gratitude toward our friends in the Bureau. . . ."

245. *"So, feeling thus":* National Archives: General Correspondence. Letter, William Alexander to chief of Weather Bureau, Nov. 20, 1900.

246. *At 11:30 A.M., Joseph:* National Archives: General Correspondence. Telegram, Joseph Cline to Weather Bureau, Sept. 11, 1900.

246. *Exactly three minutes later:* National Archives: General Correspondence. Telegram, Isaac Cline to Weather Bureau, Sept. 11, 1900.

247. *"I wish to report":* Galveston *News,* Sept. 17, 1900.

247. *Isaac could not help it:* Isaac never directly states that he should have taken his family to the Levy Building early on, but how could any man in a similar position avoid such thoughts?

247. *Joseph, underneath:* See Joseph's memoir, *When the Heavens Frowned.* In his chapter on the Galveston hurricane, 49–63, Joseph clearly, if at times obliquely, claims to have recognized the true danger of the storm when Isaac did not. See, for example, page 53, when he writes, "Until my statement of the danger, everyone there had believed [Isaac's house] to be immune to destruction by storm." Everyone, presumably, including his brother.

248. *There were dreams:* I base this observation on human nature. What survivor of a tragedy has never dreamed that the outcome had been different?

248. *"A dream," Freud wrote:* Freud, 155.

248. *"The hurricane which visited":* Cline, "Special Report," 372.

248. *"Storm warnings were timely":* Ibid., 373.

248. *"As a result thousands":* Ibid., 373.

249. *In later years:* Letter, E. M. Vernon, chief, Forecasts and Synoptic Reports Division, to M. S. Douglas, Nov. 9, 1956. "It is estimated," Isaac Cline wrote, "that about 12,000 people moved out prior to the crisis, otherwise the loss of life would doubtless have been more than double what it was. . . ." Isaac deployed the passive voice whenever he sought to describe something he himself had done. The 12,000 is almost certainly his own estimate. Rosenberg Library. 95-00020. Box 1, File 7.

249. *"Among the lost":* Cline, "Special Report," 373.

249. *"My personal experience":* National Archives: General Correspondence. Letter, I. M. Cline to chief of the Weather Bureau, Sept. 23, 1900. Box 1476.

Washington: A Letter from Moore

250. *"The practical inutility":* Houston *Post,* Sept. 14, 1900.

250. *Moore, in a five-page letter:* National Archives: General Correspondence. Draft of letter, Willis Moore to Houston *Post,* Sept. 22, 1900. Letter as published, Houston *Post,* Sept. 28, 1900. Box 1476.

251. *"We would all rather believe":* Houston *Post,* Sept. 28, 1900.

251. *Isaac mailed the clippings:* National Archives: General Correspondence. Letter, I. M. Cline to chief of the Weather Bureau, Sept. 28, 1900. Box 1476.

251. *"Regarding the warnings":* Ibid., 2.

252. *"If I had taken the time":* Cline, *Storms,* 98.

252. *"I did not foresee":* Ibid., 99.

252. *The Boston* Herald: *Monthly Weather Review,* Sept. 1900, 376.

252. *The Buffalo, New York,* Courier: Ibid., 376.

253. *The* Inter-Ocean: Ibid., 377.

253. *"An example?":* National Archives: General Correspondence. See clipping from *Diario de la Marina,* Sept. 18, 1900. Box 1475.

254. *"It is apparent to me:* National Archives: General Correspondence. Letter, Willis Moore to secretary of agriculture, Sept. 21, 1900. Box 1475.

28th and P: The Ring

255. *There were miracles:* First Baptist Church, 7.

255. *"The dreams of young children":* Freud, 160.

256. *Isaac, Joseph, and John Blagden:* Circular. Office of chief clerk. Weather Bureau. Sept. 28, 1900.

256. *Someone donated a case:* Barton. Report of Red Cross Relief, Galveston, Texas, 52.

256. *The* Palmetto Post: *Palmetto Post*, Sept. 20, 1900. In Barton.

256. *"It is," she wrote:* Barton. Letter, Clara Barton to "the Public," Oct. 6, 1900.

257. *Toward evening:* Weems, 146. Weems states the crew found Cora's body shortly after nightfall, indicating the crew had begun working in that area some time earlier.

257. *"Even in death":* Cline, "Cyclones," 15.

258. *The body was transported:* Lakeview Cemetery Record. Vol. I. 1992. Rosenberg Library.

258. *Isaac kept the ring:* Isaac nowhere states this. It is conjecture, purely, but I base it on a number of things, particularly: Isaac's essentially romantic character; his devotion to Cora; his deep knowledge of portraiture and the symbolic messages embedded within by their painters; the fact he wears a diamond in Whitesell's photographic portrait (see note for page 4, *A New Orleans photographer*); and Whitesell's obvious effort to light the ring and Isaac's eyes so that both gleam from the darkness.

258. *That night:* West. Chronology.

PART VI: HAUNTED

Isaac: Haunted

263. *On Monday, September 10, Willis Moore:* National Archives: General Correspondence. Telegram, Sept. 10, 1900, Willis L. Moore to City Editor, *Evening World*. The initials H. C. F. beneath Moore's name suggest that the content of the telegram, perhaps the telegram itself, actually came from H. C. Frankenfield, one of the bureau's senior scientists. I cite Frankenfield's reminiscence about weather school in a note for page 33 (*"You will cheerfully"*).

263. *It rapidly regained power:* National Archives: General Correspondence. See exchange of letters, beginning Sept. 13, 1900, A. I. Root to U.S. Weather Bureau. Box 1476.

263. *The Central Office:* Ibid., Sept. 28, 1900.

263. *It killed six loggers:* "Six Drowned in Wisconsin." Untitled dispatch in Rosenberg Library. Storm of 1900 Collection. Subject File. News Clippings. The item is dated Sept. 15, 1900.

264. *Manhattan, half a continent:* Fernandez-Partagas, 105, note 63.

264. *The storm sank:* Ibid., 106, notes 66–71.

264. *The city conducted:* Mart H. Royston Papers. Oct. 9, 1900. Rosenberg Library. Manuscript Collection. 25-0587.

264. *Early in 1901:* McComb, 122.

265. *"Many people":* Personal Accounts: Cortes, 6

265. *They created:* World's Fair Bulletin, April 1904, 24–30. Rosenberg Library. 76-0004.

265. *It rose seventeen feet:* Personal Accounts: Stuart, 78.

265. McClure's Magazine *called it:* Turner, George, 615.

265. *They raised the altitude:* Cartwright, *Galveston,* 29.

265. *The city built:* Ibid., 29.

265. *To signal the city's faith:* Ibid., 5.

266. *"We have got down":* "Chicago," 686.

266. *Just four months later:* Kane, 171–73.

267. *Soon after the storm:* National Archives: Letters Sent. Letter, Willis Moore to secretary of agriculture, Oct. 9, 1900.

267. *On November 3, 1900:* National Archives: General Correspondence. Letter, I. M. Cline to chief of Weather Bureau, Nov. 3, 1900.

267. *"I believe that I have":* National Archives: General Correspondence. Letter, J. L. Cline to chief of Weather Bureau, Dec. 4, 1900.

267. *By then, however:* National Archives: Letters Sent. Letter, Willis Moore to secretary of agriculture, April 5, 1901.

267. *Two weeks later:* National Archives: Letters Sent. Letter, Willis Moore to secretary of agriculture, April 20, 1901.

268. *In 1909:* Whitnah, 122.

268. *"When a station official":* Cline, *Storms,* 140.

269. *"The object":* Ibid., 141.

269. *Isaac's disillusionment.:* Ibid., 141–46.

269. *At nine o'clock:* Ibid., 146.

269. *The clearest evidence:* National Archives. Letter, March 30, 1922, Joseph L. Cline to Henry E. Williams. Reminiscences of Employees. Other Records, 1878–1924. Box 1.

270. *He gave up the study:* Cline, "Century," 37.

270. *In his monograph:* Ibid., 37.

271. *In September 1909: Monthly Weather Review.* Sept. 1909, 625.

271. *On the morning of:* First Baptist Church, 8.

271. *A month later:* Ibid., 8.

271. *He divided his annual leave:* Cline, *Storms,* 127.

271. *He collected:* Ibid.,170.

272. *"Time lost can never be recovered":* Ibid., 248.

272. *Isaac Monroe Cline:* Weems, 164.

272. *"Galveston should take heart":* National Archives: General Correspondence. Telegram, Willis Moore to Chicago *Tribune,* Sept. 13, 1900.

272. *The seawall held:* McComb, 149.

272. *The death toll:* Author's conversations with Hugh Willoughby, Chris Landsea, Jerry Jarrell, Bill Gray, and others. See note on pages 279–80, at "The Storm: Somewhere a Butterfly."

272. *None believed:* Author's conversations, as in preceding note for page 272.

273. *There was talk:* See Emanuel, "Dependence," "Hypercanes," and "Toward."

273. *The Army Corps of Engineers:* See U.S. Army Corps of Engineers. Interim Technical Data Report. Metro New York Hurricane Transportation Study. November 1995.

273. *But in the narrow:* Author's observations.

273. *Once, in a time long past:* A small marker noting the orphanage disaster stands at the seaward rim of the seawall, opposite the Wal-Mart.

SOURCES

Abbe, Cleveland. Papers. Library of Congress.

Acheson, Sam. *35,000 Days in Texas*. 1938. Reprint, Westport, Conn.: Greenwood Press, 1973.

Alexander, W. H. "Thunderstorms at Antigua, W.I." *Monthly Weather Review*. U.S. Weather Bureau, Washington, D.C., September 1900.

Anthes, Richard A. "Tropical Cyclones: Their Evolution, Structure and Effects." *Meteorological Monographs*. American Meteorological Society, vol. 19, no. 41, February 1982.

Barton, Clara. Papers. Library of Congress.

Bigelow, Frank H. "Report on the Temperatures and Vapor Tensions of the United States." Bulletin S. Weather Bureau. 1906.

Birch, Doug. "The Incredible Shrinking Glacier," *Baltimore Sun*, February 10, 1997, 2a.

Cartwright, Gary. "The Big Blow." *Texas Monthly*, August 1990, 76–81.

———. *Galveston*. New York: Atheneum, 1991.

"Chicago and Galveston." *McClure's Magazine*, April 1907, 685–86.

City Directory, Galveston. 1899–1900. Rosenberg Library.

Cline, Isaac M. "Address of Dr. Cline, Delivered to the YMCA Saturday Night." *Galveston News*, December 21, 1891.

———. "A Century of Progress in the Study of Cyclones." President's Address, American Meteorological Society, December 29, 1934. Published, New Orleans, 1942.

———. "Cyclones, Hurricanes and Typhoons and Other Storms." Speech, Isaac Delgado Central Trades School, October 2, 1936.

———. "Relation of Changes in Storm Tides on the Coast of the Gulf of Mexico to the Center and Movement of Hurricanes." Proceedings of the Louisiana Engineering Society, New Orleans, La., vol. 6, no. 5, October 1920.

———. "Special Report on the Galveston Hurricane of September 8, 1900." *Monthly Weather Review*. U.S. Weather Bureau. Washington, D.C., November 16, 1900, 372–374.

———. *Storms, Floods and Sunshine*. New Orleans: Pelican Press, 1945.

———. "Summer Hot Winds on the Great Plains." Address to Philosophical Society of Washington, January 20, 1894. Published by the Society, March 1894.

———. *Tropical Cyclones*. New York: Macmillan, 1926.

———. "West India Hurricanes." Galveston *News*, July 16, 1891.

Cline, Joseph L. *When the Heavens Frowned*. Dallas: Mathis, Van Nort & Co., 1946.

Coulter, John, ed. *The Complete Story of the Galveston Horror*. United Publishers of America, 1900.

Daily Journal, Galveston Station. U.S. Weather Bureau, Department of Agriculture, September 1900. Rosenberg Library.

Douglas, Marjory Stoneman. *Hurricane*. New York: Rinehart & Co., 1958.

Dunn, Gordon E., and Banner I. Miller. *Atlantic Hurricanes*. Baton Rouge, La.: Louisiana State University Press, 1960.

Eisenhour, Virginia. *The Strand of Galveston*. 1973. Rosenberg Library, Galveston, Texas.

Emanuel, Kerry A. "The Dependence of Hurricane Intensity on Climate." *Nature*, vol. 326, no. 2, April 1987, 483–85

———. "Hypercanes: A Possible Link in Global Extinction Scenarios." *Journal of Geophysical Research*, vol. 100, no. D7, July 20, 1995, 13755–65.

———. "Toward a General Theory of Hurricanes." *American Scientist*, vol. 76, July–August 1988, 371–79

Fernandez-Partagas, Jose. Unpublished manuscript. National Hurricane Center, National Oceanic and Atmospheric Administration, Miami.

Fire Insurance Map. Galveston. New York: Sanborn-Perris Map Co., 1899.

First Baptist Church. Account. Rosenberg Library, Galveston, Texas. 95-0002. Box 1, Folder 7.

Frankenfield, H. C. "Some Reminiscences." National Archives. Weather Bureau. Other Records, 1878–1924.

Freud, Sigmund. *The Interpretation of Dreams*. 1900. Reprint, New York: Avon, 1965.

Friendly, Alfred. *Beaufort of the Admiralty*. New York: Random House, 1977.

Frisinger, H. Howard. *The History of Meteorology: To 1800*. New York: Science History Publications, 1977.

"The Galveston Horror," *The Western World and American Club Woman Illustrated*, Chicago, October 1900.

Garriott, E. B. "Forecasts and Warnings." *Monthly Weather Review*, U.S. Weather Bureau, Washington, D.C., August 1900.

———. "The West Indian Hurricane of September 1–12, 1900." *National Geographic*, vol. 11, no. 10, October 1900, 384–92.

Geddings, R. M. "Reminiscence." National Archives. Weather Bureau. Other Records, 1878–1924.

The Giles Mercantile Agency Reference Book. Galveston, Texas, 1899.

Gray, William M. "Strong Association Between West African Rainfall and U.S. Land-fall of Intense Hurricanes." *Science*, vol. 249, September 14, 1990, 1251–56.

Gray, William M., John D. Sheaffer, and Christopher W. Landsea. "Two Climate Trends Associated with Multidecadal Variability of Atlantic Hurricane Activity." In *Hurricanes*, by Henry F. Diaz and Roger S. Pulwarty. New York: Springer-Verlag, 1997.

Greely, A. W. "Hurricanes on the Coast of Texas." *National Geographic*, vol. 11, no. 11, November 1900, 442–45.

Gregg, Josiah. *Diary and Letters of Josiah Gregg. Southwestern Enterprises. 1840–1847.* Norman, Okla.: University of Oklahoma Press, 1941.

Henry, Walter K., Dennis M. Driscoll, and J. Patrick McCormack. "Hurricanes on the Texas Coast." Center for Applied Geosciences, Texas A&M University, May 1982.

Hughes, Patrick. *American Weather Stories*. Washington, D.C.: U.S. Department of Commerce, 1976

International Cloud Atlas. Vol. 2. World Meteorological Organization, 1987.

Kane, Harnett T. *The Golden Coast*. Garden City, N.Y.: Doubleday, 1959.

Landsea, Christopher W., William M. Gray, Paul W. Mielke Jr., and Kenneth J. Berry. "Seasonal Forecasting of Atlantic Hurricane Activity." *Weather*, vol. 49, no. 8, August 1994, 273–84.

Laskin, David. *Braving the Elements*. New York: Anchor-Doubleday, 1996.

Liu, G., J. A. Curry, and C. A. Clayson. "Study of Tropical Cyclogenesis Using Satellite Data." *Meteorology and Atmospheric Physics*, vol. 56, 1995, 111–23.

Lockhart, Gary. *The Weather Companion*. New York: John Wiley & Sons, 1988.

Ludlum, David M. *Early American Hurricanes, 1492–1870*. Boston: American Meteorological Society, 1963.

Mason, Herbert Molloy. *Death from the Sea.* New York: Dial, 1972.

McComb, David G. *Galveston: A History.* Austin, Texas: University of Texas Press, 1986.

McCullough, David. *Mornings on Horseback.* New York: Simon & Schuster, 1981.

———. *The Path Between the Seas.* New York: Simon & Schuster, 1977.

McDonough, P. "The Barbados Hurricane of August 10th, 1831." Manuscript, March 24, 1900. National Archives. General Correspondence. Box 1445.

Montgomery, Michael T., and Brian F. Farrell. "Tropical Cyclone Formation." *Journal of Atmospheric Sciences*, vol. 50, no. 2, January 1993, 285–310.

Monthly Weather Review. U.S. Weather Bureau. Washington, D.C.

Morison, Samuel Eliot. *Admiral of the Ocean Sea.* Boston: Little, Brown & Co., 1942.

Morris, Sylvia Stallings, ed. *William Marsh Rice and His Institute.* Houston: Rice University, 1972.

Nathan, Anne, and Harry I. Cohen. *The Man Who Stayed in Texas.* New York: McGraw-Hill, 1941.

National Archives. Department of Agriculture. Weather Bureau.

— Administrative and Fiscal Records. Miscellaneous Operational Records, 1871–1912: General Orders, Circulars, and Instructions, c. 1871–1909.

— General Correspondence. Letters Received. 1894–1911.

— Inspection Reports.

— Letters Sent by Chief of the Bureau, 1891–1895, 1897–1911. vol. 9.

— Records of Surface Land Observations.

Observations. Handwritten log of weather, U.S. Weather Bureau, Galveston. Rosenberg Library, Galveston, Texas. Storm of 1900 Collection. 85-0020.

Ousley, Clarence. *Galveston in Nineteen Hundred.* Atlanta: William C. Chase, 1900.

Personal Accounts. Rosenberg Library, Galveston, Texas.

— Beal, Emma. Interview with Marilee Neale. Oral History Project. Part 1, April 26, 1972. Part 2, June 14, 1972.

— Benjamin, David. "From the Stricken City." Undated news clipping. Storm of 1900 Collection. Subject File. Newspaper Clippings.

— Bettencourt, Mr. and Mrs. Henry. Interview. Oral History Project.

— Blagden, John. Letter. September 10, 1900. Manuscript Collection. 46-0006.

— "Charlie." Letter to Mrs. Law. September 12, 1900. Storm of 1900 Collection. Storm Letter. 91-0012B.

—Cortes, Henry C. "Personal Recollections of the 1900 Galveston Hurricane." William Maury Darst Papers. 93-0023. Box 42, File 15.

—Davis, Walter. Letter to mother. September 14, 1900. Manuscript Collection. 85-0018B.

—Deer, J. N. Letter, Deer to R. N. Schooling. September 22, 1900. Storm of 1900 Collection. Letters and Memoirs.

—Focke, Mrs. John. "September Eighth 1900." Letter to daughters. Manuscript Collection. 67-0006

—Goodman, R. Wilbur. "Address on The Great Galveston Storm of 1900." August 13, 1975. William Maury Darst Papers. 93-0023. Box 14, File 17.

—Harris, John W. Interview with Robert L. Jones. Oral History Project. December 22, 1980.

—Hawley, J. H. Letter, Hawley to wife and daughters. September 18, 1900. Manuscript Collection. 67-0042B.

—Hawley, Sarah Davis. Letter, Hawley to mother. September 12, 1900. Manuscript Collection, 97-0020. Box 1, File 14.

—Hopkins, Louise. "The Day I Can't Forget." Interview with Jane Kenamore. Storm of 1900 Subject File. Map of Storm Damage/Pictures.

—Hopkins, Louise. Interview. Oral History Project. July 8, 1982.

—Lang, Marie Berryman. "Mother Nature Turns Fury on Galveston." From *East Texas Catholic*. Storm of 1900 Collection. Subject File. Newspaper Clippings.

—Lewis, Wilber. Letter, Lewis to "Dear Friend." September 24, 1900. Storm of 1900 Collection. Letters and Memoirs. Subject File.

—Monagan, Tom L. "Has Relics of 1900 Storm." Dallas *Morning News,* June 7, 1925. In J. L. Monagan Papers. 28-0175.

—Sterett, William. "The Scene Coming into Galveston after the Storm Can Not Be Adequately Described." Storm of 1900 Collection. Subject File. Newspaper Clippings.

—Stuart, Ben C. "Storms." Ben C. Stuart Papers. 29-0147. Box 1.

—Tipp, Phillip Gordie. "Memories of September the 8th 1900 at Galveston, Texas." 05-0002. Box 1, File 7.

—Wolfram, A. R. "The Galveston Storm of 1900." William Maury Darst Papers. 93-0023. Box 14, File 13.

"Picturesque Galveston." Galveston *Tribune*. Possibly 1899. Rosenberg Library, Galveston, Texas.

Piddington, Henry. *The Sailor's Horn-Book for the Law of Storms*. 6th ed. London: Frederic Norgate, 1876.

Pielke, Roger A., Sr. *The Hurricane*. London: Routledge, 1990.

Pielke, Roger, Jr., and Roger A. Pielke Sr. *Hurricanes: Their Nature and Impacts on Society*. Chichester, England: John Wiley & Sons, 1997.

Rappaport, Edward N., and Jose Fernandez-Partagas. "The Deadliest Atlantic Tropical Cyclones, 1492 to Present." National Hurricane Center, National Oceanic and Atmospheric Administration, Miami. Updated, June 30, 1998.

Reid, Lt. Col. W. *An Attempt to Develop The Law of Storms*. London: John Weale, 1846.

Report of the Central Relief Committee for Galveston Storm Sufferers. Galveston. May 2, 1902. Rosenberg Library, Galveston, Texas.

Rollfing, Louisa Christina. *Autobiography*. Unpublished manuscript, vols. 1–5. Rosenberg Library, Galveston, Texas. Manuscript Collection. 83-0054.

Rosenberg Library. Galveston and Texas History Center. Galveston, Texas.

Rosser, Angie Ousley. "In the Eye of Galveston's Great Storm." Houston *Chronicle,* September 8, 1980.

Rosser, W. H. *The Law of Storms*. London: Norie & Wilson, 1886.

St. Mary's Hospital. *A Pattern of Love*. Commemorative booklet. Rosenberg Library.

Simpson, R. H. "Hurricanes." *Scientific American*, vol. 190, no. 6, June 1954, 32–37.

Snow, Edward Rowe. *Astounding Tales of the Sea*. New York: Dodd, Mead & Co., 1965.

Tannehill, Ivan Ray. *Hurricanes*. Princeton, N.J.: Princeton University Press, 1942.

Tapp, Ruby (Credo). "The Great Galveston Storm." *Tempo Magazine,* Miami, September 8, 1968, 8–12.

Thomas, R. *Interesting and Authentic Narratives of the Most Remarkable Shipwrecks, Fires, Famines, Calamities, Providential Deliverances, and Lamentable Disasters on the Seas, in Most Parts of the World*. Hartford, Conn.: Ezra Strong, 1839.

Traxel, David. *1898*. New York: Alfred A. Knopf, 1998.

Trefil, James. *Meditations at Sunset*. New York: Charles Scribner's Sons, 1987.

Turner, Elizabeth Hayes. *Women, Culture, and Community: Religion and Reform in Galveston, 1880–1920*. Oxford: Oxford University Press, 1997.

Turner, George Kibbe. "Galveston: A Business Corporation." *McClure's Magazine,* October 1906, 610–620.

Verne, Jules. *Twenty Thousand Leagues Under the Sea*. Edited by Walter James Miller and Frederick Paul Walter. Annapolis, Md.: Naval Institute Press, 1993.

Volland, Hans, ed. *CRC Handbook of Atmospherics*. Vol. 1. Boca Raton, Fla.: CRC Press, 1982.

von Herrmann, Charles F. Reminiscence, April 17, 1922. National Archives. Weather Bureau. Other Records, 1878–1924.

Watson, Lyall. *Heaven's Breath*. New York: William Morrow, 1984.

Weems, John Edward. *A Weekend in September*. 1957. Reprint, College Station, Texas: Texas A&M University Press, 1980.

West, Rebecca. *1900*. London: Weidenfeld & Nicolson, 1996.

Whitnah, Donald R. *A History of the United States Weather Bureau*. Urbana, Ill.: University of Illinois Press, 1961.

Young, S. O. "Interesting Account of the Great Hurricane." Rosenberg Library. Storm of 1900 Collection. Subject File. News Clips.

Zebrowski, Ernest, Jr. *Perils of a Restless Planet*. Cambridge: Cambridge University Press, 1997.

ACKNOWLEDGMENTS

NO BOOK CAN be written without the help of legions of good souls who donate their time and energy to the author's cause. I wish first to thank my wife, Christine Gleason, a natural editor who happens also to be a brilliant physician. Her repeated readings of the manuscript and her observations were invaluable, even the occasional *Zzzzzz* she wrote in the margins to note places where the story dragged. My daughters, Kristen, Lauren, and Erin, showed unusual patience in tolerating their father's predawn disappearances into his office and his mysterious ban on nonessential use of his computer, a prohibition based entirely on his neurotic fear that something catastrophic might befall his manuscript.

My agent, David Black, is that rarest of agents who insists that a book proposal be just exactly right, even in the face of death threats from his authors. He is an excellent human being with an unerring eye for story. My editor, Betty Prashker, was as always a cheerfully assertive voice, prodding me gently for the manuscript and, later, recommending with equal grace that parts of it be destroyed and never seen by any other human reader.

The observations of friends who read all or parts of the manuscript were invaluable. Thanks, then, to Robin Marantz-Henig, who read the entire thing and gave me a detailed structural and stylistic critique, and to Alex Kotlowitz and Carrie Dolan, whose encouragement helped me survive those dark early days when the writing first got under way. I owe a special debt to Hugh E. Willoughby, director of the Hurricane

Research Division of NOAA's Atlantic Oceanographic and Meteorological Laboratory, Virginia Key, Florida, who showed himself to be a deft critic of style as well as meteorological content. Any lingering errors are entirely my fault, not his.

Many archivists and librarians helped make my journey back to Isaac's time a pleasant one, foremost among them Casey Greene, head of Special Collections, Shelly Henly Kelly, and Anna B. Peebler, all of the Rosenberg Library. Special thanks to Casey for no doubt saving my life with his frequent warnings not to tip my chair so far back. I thank, too, Margaret Doran, Curator of Collections at Galveston's Moody Museum, for showing me the letters of lovestruck Will Moody Jr. and his beloved "Hib," neither of whom made it into the final draft but whose passion and observations nonetheless inform the narrative. Thanks also to the Seattle Public Library, for making available to me a berth in the Writer's Room and to everyone at the National Archives Annex, College Park, Maryland, the Library of Congress, and the Suzallo Library of the University of Washington—just for being there and for helping to preserve so many bits and pieces of the nation's past and for happily fielding endless, naive queries about times long gone.

Erik Larson
Seattle
1999

Index